FLUID MECHANICS AND INDUSTRIAL APPLICATIONS

The book captures the essence of the Eighth International Conference on Fluid Mechanics and Industrial Applications (FMIA 2025). These original and peer-reviewed research papers delve into recent developments and applications of parallel computing in the field of fluid mechanics, CFD, fluid power, fluid flow and related disciplines.

This volume covers basic formulations and their computer modelling as well as the relationship between experimental and analytical results. It touches upon a wide array of topics including Aerodynamics, Thermodynamics and Heat Transfer, and Fluid Theory and Related Algorithms. The inclusion of innovative research contributions and practical applications makes this book a repository of novel ideas highlighting the latest developments in fluid mechanics.

This is a highly beneficial guide for researchers from industry and academia, practitioners, students, and technology enthusiasts.

PROCEEDINGS OF THE EIGHTH INTERNATIONAL CONFERENCE ON FLUID MECHANICS AND INDUSTRIAL APPLICATIONS (FMIA 2025), WUHAN, CHINA, NOVEMBER 15TH–16TH, 2025

Fluid Mechanics and Industrial Applications

Edited by

Zhigang Fang
Wuhan University of Technology, China

CRC Press
Taylor & Francis Group
Boca Raton London New York Leiden

CRC Press is an imprint of the
Taylor & Francis Group, an **informa** business

First edition published 2026
by CRC Press
4 Park Square, Milton Park, Abingdon, Oxon, OX14 4RN

and by CRC Press
2385 NW Executive Center Drive, Suite 320, Boca Raton FL 33431

British Library Cataloguing-in-Publication Data
A catalogue record for this book is available from the British Library

ISBN: 978-1-041-15371-9 (hbk)
ISBN: 978-1-041-15463-1 (pbk)
ISBN: 978-1-003-67956-1 (ebk)

DOI: 10.1201/9781003679561

Typeset in Times New Roman
by MPS Limited, Chennai, India

Fluid Mechanics and Industrial Applications – Fang (Ed.)
© 2026 The Editor, ISBN: 978-1-041-15371-9

Contents

Fluid Mechanics and Industrial Applications – Fang (Ed.)
© 2026 The Editor, ISBN: 978-1-041-15371-9

Preface

2025 the 8th International Conference on Fluid Mechanics and Industrial Applications (FMIA 2025) included original and peer-reviewed research papers and held on November 15–16, 2025, in Taiyuan, China. This conference was sponsored by Asian Union of Information Technology, and was attended by academicians, researchers, entrepreneurs, government agencies and policy-makers, postgraduates.

FMIA is an international conference devoted to the discussion of recent developments and applications of parallel computing in the field of fluid mechanics, CFD, fluid power, fluid flow and related disciplines. The field of fluid mechanics is vast and covers numerous and diverse applications. The conference covered a wide range of topics, including basic formulations and their computer modelling as well as the relationship between experimental and analytical results. The meeting provided a forum for discussing new work on fluid mechanics and in particular for promoting the interchange of novel ideas and the presentations of the latest developments in this field. This conference would also provide an ideal environment to develop new collaborations and meet experts on the fundamentals, applications, and products of the mentioned fields.

This two-day conference was organized in technical association with the Asian Union of Information Technology and included invited keynote talks and oral paper presentations from both academia and industry. The topics covered in this FMIA 2025 are: (1) Aerodynamics; (2) Thermodynamics and Heat Transfer; (3) Fluid Theory and Related Algorithms. Each keynote speakers lasted for 30 minutes, including 5 minutes for Q&A. and for other sessions, there was 20 minutes for each presentation, including 5 minutes for Q&A. The book is useful to professionals and scientists involved or interested in fluid mechanics and industrial applications. Hopefully, all participants and other interested readers would benefit scientifically from the proceedings and also find it stimulating in the process.

We would like to thank all the author submitted papers to this FMIA 2025 and thank all the reviewers for their time and effort in reviewing articles. Especially we would like to thank the organizing committee for their valuable advices in the organization and helpful peer review of the papers.

Conference Organizing Committee
Wuhan, China

Preface

Fluid Mechanics and Industrial Applications – Fang (Ed.)
© 2026 The Editor, ISBN: 978-1-041-15371-9

About the editor

Zhigang Fang received his Ph.D. in 2013 and joined Wuhan University of Technology as a lecturer in 2014. His research focuses on energy recovery of automobile shock absorbers, vehicle energy-saving technology, CFD, heat transfer, energy conversion, also engaged in the modelling and simulation of complex systems in multidisciplinary fields, hydraulic and fluid mechanics simulation, system dynamics simulation, energy conversion and efficiency simulation and verification.

Fluid Mechanics and Industrial Applications – Fang (Ed.)
© 2026 The Editor, ISBN: 978-1-041-15371-9

Committees

Chairmen
Prof. Wei Zhang, *Hunan University, China*
Prof. Jianjun Xu, *Northeast Petroleum University, China*

Scientific committee
Dr. Yong Wang, *Chongqing University, China*
Prof. John Herlihy, *Merced College, USA*
Associate Prof. S. Zhu, *Wuhan University, China*
Dr. Bian Gong, *Shenzhen Technology University, Shenzhen, China*
Dr. Xintian Zhou, *Jiangsu Ocean University, China*
Dr. Yufei He, *Hunan Institute of Engineering, China*
Dr. Lei Jiang, *Nanjing University of Science and Technology, China*
Dr. Buyan Yu, *Changsha University, China*
Dr. Qifei Du, *Tiangong University, Tianjin, China*
Dr. Jing Cai, *Dongfeng Motor Corporation Technical Center, China*
Dr. D.M. Li, *Shandong University of Science and Technology, China*
Dr. Han Zhang, *Guangzhou Automobile Group Co., Ltd., China*
Dr. Xiaohe Deng, *Wuhan University of Technology, China*
Dr. Zhibin You, *Wuhan University of Technology, China*
Dr. Jun Xiao, *Wuhan University of Technology, China*
Dr. Zhihui Tan, *Wuhan University of Technology, China*

Organizing committee
Dr. Jun Xiao, *Wuhan University of Technology, China*
Dr. Zhihui Tan, *Wuhan University of Technology, China*
Dr. Zhigang Fang, *Wuhan University of Technology, China*
Dr. Shan Zhu, *Northeast Petroleum University, China*
Dr. Long Chen, *Northeast Petroleum University, China*
Mr. Wu, *Wuhan Xuerong Conference Services Co., Ltd., China*
Mr. Ke, *Wuhan Xuerong Conference Services Co., Ltd., China*
Mr. Ye, *Wuhan Xuerong Conference Services Co., Ltd., China*
Mr. Bao, *Wuhan Xuerong Conference Services Co., Ltd., China*
Mr. Xiao, *Wuhan Xuerong Conference Services Co., Ltd., China*

Fluid Mechanics and Industrial Applications – Fang (Ed.)
© 2026 The Author, ISBN: 978-1-041-15371-9

Study on flow instability of supercritical RP-3 in a multi-channel system

Jingsong Huang
National Key Laboratory of Science and Technology on Aero-Engine Aero-Thermodynamics, School of Energy and Power Engineering, Beihang University, Beijing, China

Zeyuan Cheng
National Key Laboratory of Science and Technology on Aero-Engine Aero-Thermodynamics, School of Energy and Power Engineering, Beihang University, Beijing, China
National Key Laboratory of Ramjet, Beijing Power Machinery Institute, Beijing, China

Yaoxian Jiao and Jianqin Zhu
National Key Laboratory of Science and Technology on Aero-Engine Aero-Thermodynamics, School of Energy and Power Engineering, Beihang University, Beijing, China

ABSTRACT: Regenerative cooling technology is a critical solution to ensure the reliable operation of scramjet combustion chambers under extreme thermal environments. However, drastic variations in fuel properties can trigger flow instability, risking local cooling failure and flight safety. The interaction and propagation mechanisms of such instabilities in multi-channel systems remain poorly understood. In this study, the regenerative cooling system is simplified into a multi-channel configuration, and a three-dimensional numerical simulation is conducted using supercritical RP-3 as the coolant. The results reveal that, under the gravity, flow instability initiates in low-flow channels, followed by delayed instability in high-flow channels, eventually forming density wave oscillations. The initial instability stems from strong density gradients in low-flow regions, which induce downstream turbulence. This increases resistance in adjacent channels, triggering a positive feedback loop involving flow rate, fluid density, and pressure drop. Consequently, instability spreads, evolving into a self-excited, sustained oscillation across the system.

1 INSTRUCTION

In hypersonic flight, the extreme thermal environment in scramjet combustion chambers makes regenerative cooling with supercritical hydrocarbon fuels a key strategy for thermal protection (Zhu *et al.* 2018). However, the intrinsic multi-channel configuration of regenerative cooling systems, combined with the drastic property variations of supercritical hydrocarbon fuels (Lin *et al.* 2024), can lead to flow instabilities within the cooling passages. These instabilities are typically manifested as uneven mass flow distribution among channels and oscillations in flow rate, pressure, and temperature. Such phenomena can significantly reduce the reliability of the cooling system, causing partial cooling channel failure and even local burnout due to overheating. Therefore, elucidating the flow instability mechanisms of supercritical hydrocarbon fuels in a multi-channel system is critical for the safe and reliable design of regenerative cooling systems.

In recent years, extensive studies have been conducted on flow instabilities of supercritical hydrocarbon fuels in single-channel configurations (Wang *et al.* 2024). Yan *et al.* (2018)

DOI: 10.1201/9781003679561-1

experimentally observed that turbulent transition and dramatic density variation of supercritical n-decane can induce flow instabilities. Zhou *et al.* (2014) reported that instabilities typically occur near the pseudo-critical and thermal cracking regions and proposed corresponding suppression strategies. Jin *et al.* (2023), using a one-dimensional transient numerical model, investigated how operating pressure and channel length affect instability characteristics of supercritical hydrocarbon fuels. Du *et al.* (2025) also employed a one-dimensional transient method and developed a hydrodynamic characteristic curve of supercritical n-decane, revealing that increasing the inlet loss coefficient improves the system stability.

However, flow distribution discrepancies in multi-channel regenerative cooling systems significantly affect the instability characteristics of supercritical hydrocarbon fuels. Fu *et al.* (2015) found that density variation is a key factor governing flow distribution in parallel helical tubes carrying supercritical aviation kerosene. Jing (2018) demonstrated that non-uniform heating leads to more severe flow maldistribution compared to uniform heat flux. Jin *et al.* (2024) used a one-dimensional transient model to study a four-channel system and showed that uneven flow distribution can trigger flow instability. Jiang *et al.* (2023) developed a 1D numerical model for parallel channels and indicated that flow instabilities in supercritical n-decane are prone to occur in the negative-slope regions of the hydrodynamic characteristic curve, and that increasing the number of channels further complicates the instability behavior.

In summary, although previous studies have provided valuable insights into the flow instability mechanisms of supercritical hydrocarbon fuels through experiments and one-dimensional simulations, the interaction mechanisms among sub-channels within regenerative cooling systems remain poorly understood. In this work, the regenerative cooling system is simplified into a multi-channel configuration, and three-dimensional numerical simulations are performed to investigate flow instabilities of supercritical RP-3. The objective is to elucidate the underlying mechanisms as well as the evolution and propagation of instabilities within multi-channel systems, thereby offering theoretical guidance for the safe and reliable operation of regenerative cooling in scramjet engines.

2 NUMERICAL METHODOLOGY

2.1 *Physical model and boundary conditions*

The regenerative cooling system in the scramjet engine is simplified to a multi-channel configuration with one-sided heating, as illustrated in Figure 1. The green region represents the fluid domain, and the orange region denotes the solid domain. Each square channel has a cross-section of 1.5 mm × 1.5 mm, with a 1.5 mm spacing between adjacent channels and a 0.75 mm outer wall thickness. The inlet and outlet sections are adiabatic collector chambers, each 50 mm in length, while the heating section measures 300 mm.

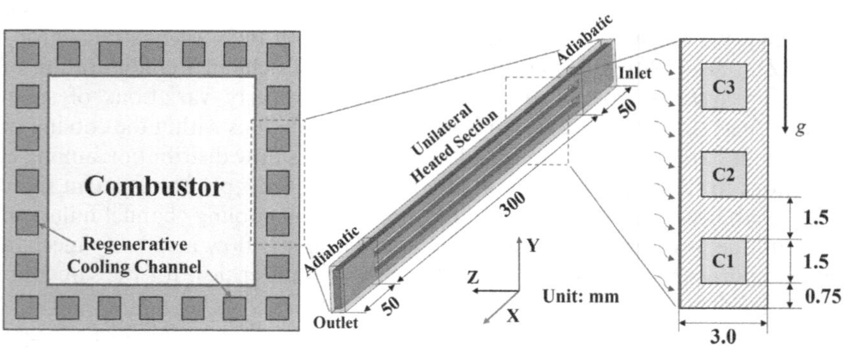

Figure 1. Schematic of the multi-channel configuration.

The fuel flows in the positive x-direction, with a no-slip condition imposed on the inner wall. A uniform heat flux is applied on the side wall in the positive z-direction, and the remaining outer walls are adiabatic. Channels are named C1, C2, and C3 from bottom to top along the positive y-direction, with gravity acting downward along the negative y-direction. The inlet mass flow rate is 3 g/s, the inlet temperature is 323 K, the heating surface heat flux is 900 kW/m^2, and the outlet pressure is set to 2.5 MPa.

2.2 *Thermophysical properties and pyrolysis model of RP-3*

RP-3 is selected as the coolant, and its thermophysical properties are calculated using a surrogate-based component model, as detailed in our previous study (Zhu *et al.* 2015). As a hydrocarbon mixture, RP-3 undergoes complex pyrolysis, producing a wide range of small alkanes, olefins, and aromatic species. This study adopts the RP-3 pyrolysis model proposed by Jiang *et al.* (2013), which involves 18 species, one primary reaction, and 23 secondary reactions. The primary reaction is:

$$C_{11.85}H_{23.82} \rightarrow 0.1086H_2 + 0.4773CH_4 + 0.5586C_2H_4 + 0.39C_2H_6 + 0.41C_3H_6$$
$$+ 0.2001C_3H_8 + 0.2246C_4H_8 + 0.0353C_4H_{10} + 0.031C_4H_6 + 0.7201C_{5+}$$
$$+ 0.27CC_{5+} + C_nH_{2n-6}$$

The pre-exponential factor for this reaction is $A = 2.869 \times 10^{14}$ 1/s, and the activation energy is $E_a = 217.9$ kJ/mol. Details of the secondary reactions can be found in Jiang *et al.* (2013).

2.3 *Numerical procedures*

Simulations are conducted using ANSYS Fluent. The governing equations are solved based on the Reynolds-Averaged Navier–Stokes (RANS) approach with the SST k-ω turbulence model. To capture flow instabilities of supercritical hydrocarbon fuels, a converged steady-state field is first obtained using the SIMPLE algorithm, followed by transient simulations with the PISO algorithm in which small disturbances are introduced at the inlet. The time step is selected to ensure a Courant number (Co) less than 10.

For discretization, gradients are computed using the least-squares cell-based method. The pressure equation uses the standard scheme, while momentum, turbulent kinetic energy, specific dissipation rate, species transport, and energy equations are discretized with second-order upwind schemes.

2.4 *Data processing method*

To facilitate the analysis of flow instability and fuel pyrolysis behavior, several dimensionless and thermophysical parameters are introduced. Their expressions are as follows:

Sub-pseudo-critical number, N_{SUBPC} (Ambrosini 2009):

$$N_{SUBPC} = \frac{\beta_{pc}}{c_{p,pc}}(h_{pc} - h_{in}) \tag{1}$$

where h_{pc} and h_{in} are the specific enthalpies at the pseudo-critical point and inlet; β_{pc} is the volume expansion coefficient; $c_{p,pc}$ is the specific heat at constant pressure.

Trans-pseudocritical number, N_{TPC} (Ambrosini 2009):

$$N_{TPC} = \frac{q_w \Pi L}{\rho_{pc} u_{in} S} \frac{\beta_{pc}}{c_{p,pc}} \tag{2}$$

where Π is the channel perimeter, S is the cross-sectional area, L is channel length, ρ_{pc} is the pseudo-critical density, and u_{in} is inlet velocity.

Average fluid temperature, T_f:

$$T_f = \frac{\int_A \rho u c_p T dA}{\int_A \rho u c_p dA} \tag{3}$$

RP-3 pyrolysis conversion rate, $Z_{RP\text{-}3}$:

$$Z_{RP-3} = \left(1 - \frac{\int_A \rho u Y_{RP-3} dA}{\int_A \rho u dA}\right) \times 100\% \tag{4}$$

where $Y_{RP\text{-}3}$ is the mass fraction of RP-3.

2.5 *Grid independence verification*

Structured grids were generated using ICEM, ensuring that all meshes satisfy $y^+ < 1$. Key monitored quantities include the outlet-averaged RP-3 temperature, pyrolysis conversion rate, and the mean relative standard deviation (MRSD) of mass flow distribution, defined as:

$$MRSD = \sum_i^n \frac{\frac{|m_i - \overline{m}|}{\overline{m}}}{n} \tag{5}$$

where m_i is the mass flow rate in channel i, $\overline{m}$ is the average flow rate across all n channels. Figure 2 shows that when the total grid number exceeds 1.2 million, the variations of all monitored parameters are less than 0.1%, indicating grid independence has been achieved.

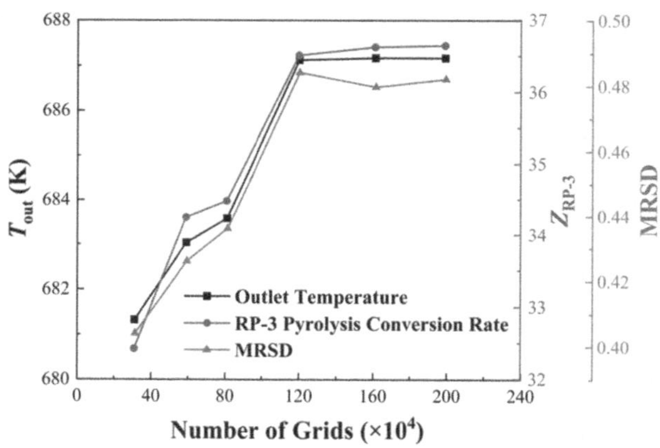

Figure 2. Grid independence verification.

2.6 *Numerical validation*

Due to the lack of experimental data on flow instability of supercritical hydrocarbon fuels including pyrolysis, a two-step validation approach is adopted. First, numerical predictions of supercritical n-decane flow instability boundaries are compared with the experimental data of Yan *et al.* (Jiang, 2013). As shown in Figure 3, the maximum relative deviation is within 5.8%. Second, simulations under the experimental conditions of Jiang *et al.* (Zhu, J.Q, 2015) are conducted. Figure 4 compares the axial distributions of the average temperature (T_f) and RP-3 pyrolysis conversion rate ($Z_{RP\text{-}3}$), with deviations less than 8%, confirming the accuracy of the numerical approach.

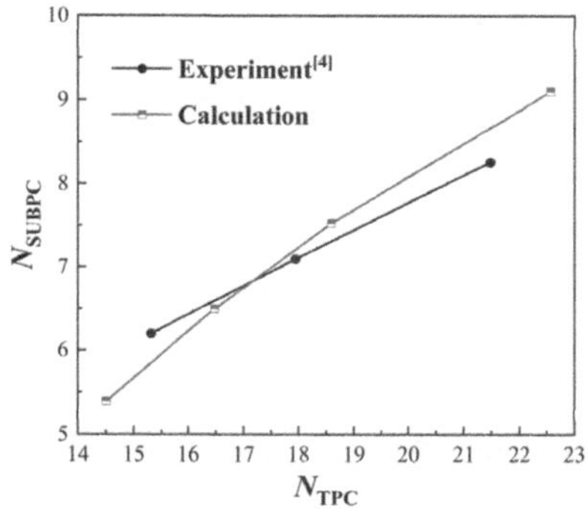

Figure 3. Validation of instability boundaries for supercritical n-decane (Jiang, 2013).

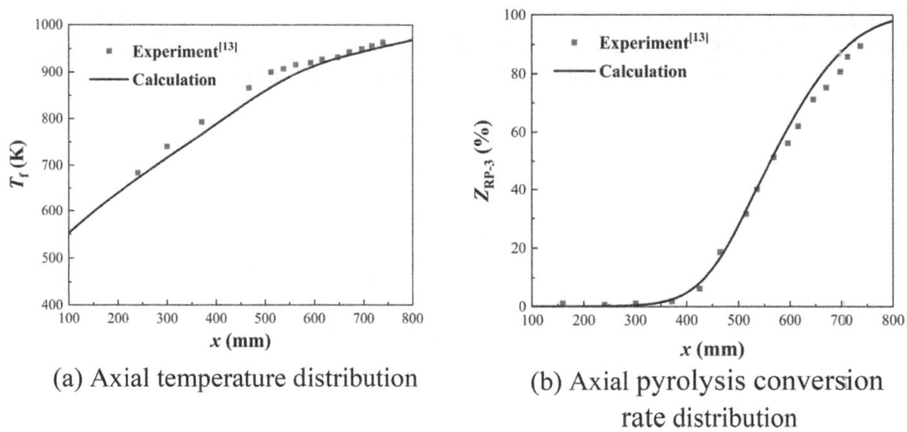

Figure 4. Validation of pyrolysis predictions (Zhu, J.Q, 2015).

3 RESULTS AND DISCUSSION

3.1 *Evolution and oscillatory characteristics of flow instability in multi-channel systems*

Figure 5 presents the temporal evolution of outlet mass flow rates in individual channels. Due to gravity along the $-y$ direction, an initial flow maldistribution is observed: C1 and C2 exhibit significantly higher flow rates than the average of 1 g/s, while the flow in C3 is only 0.4 g/s. This deviation leads to noticeable differences in flow stability among the channels. The simulation results indicate that, under initial disturbances, small-amplitude oscillations first appear in the low-flow C3 during 0–10 s, with a continuous decline in C2 and relatively stable flow in C1, implying that lower-flow channels exhibit weaker stability. Between 10–15 s, C2 shows accelerated flow decline accompanied by oscillations, while C3 experiences intensified instability with increased amplitude, and C1 begins to exhibit mild oscillations. In the 15–20 s window, C3 continues oscillating, C2 begins noticeable oscillations after 16.2 s, and C1's flow keeps decreasing. During 20–25 s, oscillations develop in C1 as well, starting around 22 s with increasing amplitude. These results suggest that the evolution of flow instability in multi-channel systems is characterized by initial low-flow channels developing small disturbances, which gradually propagate to higher-flow channels due to their parallel configuration. Eventually, this results in self-excited, sustained oscillations across all channels.

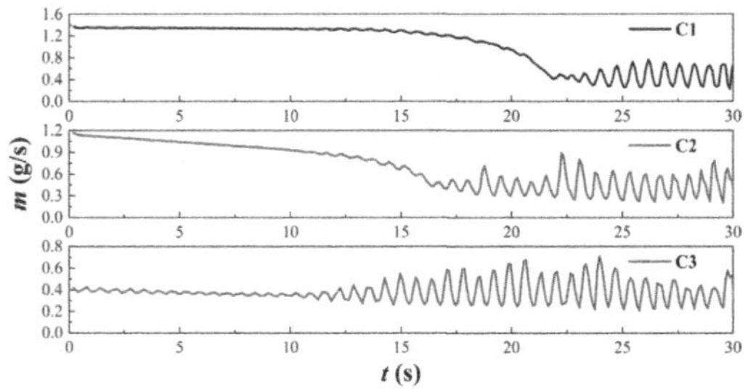

Figure 5. Temporal variation of outlet mass flow rates in each channel.

Figure 6 presents the fluctuations in outlet flow rate observed during 25–28 s. Gravity-induced maldistribution still persists, with C1 exhibiting the highest average flow rate of approximately 0.47 g/s, and C3 the lowest at around 0.35 g/s. Notably, channels with higher flow rates exhibit larger oscillation amplitudes, as C1 shows significantly greater amplitude compared to C2 and C3. In terms of period, all channels have similar oscillation frequencies; the periods for C2 and C3 are about 0.67 s, while C1's is around 0.73 s.

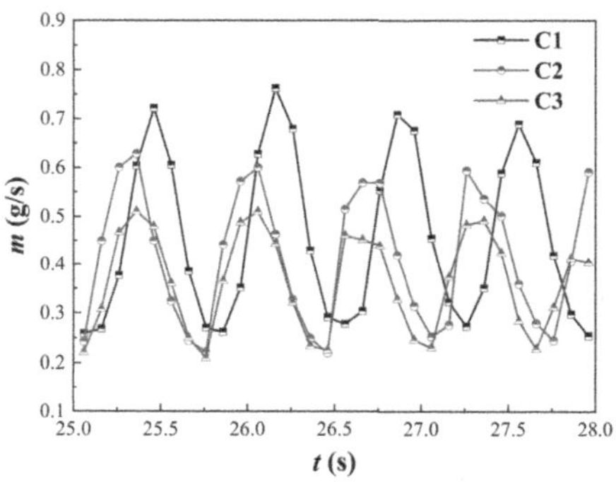

Figure 6. Outlet mass flow rate variations during 25–28 s.

Figure 7 depicts the frequency-domain characteristics of the flow oscillations. All channels show a dominant frequency $f_1 \approx 1.48$ Hz, accompanied by weak second and third harmonics. This frequency aligns with typical density-wave oscillations observed in regenerative cooling systems, suggesting that the observed oscillations are primarily of the density-wave oscillation.

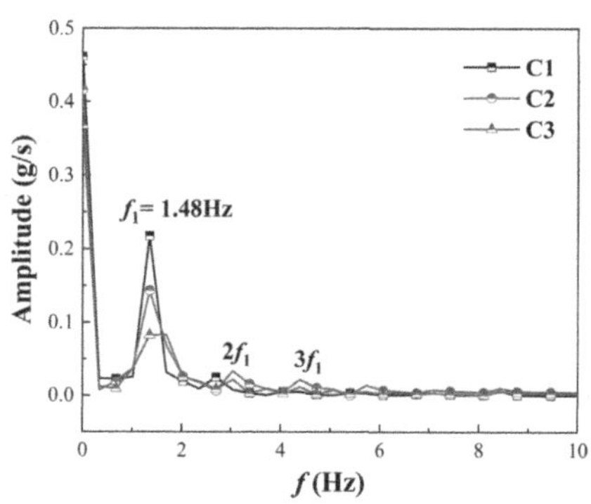

Figure 7. Frequency-domain analysis of flow rate oscillations in each channel.

3.2 *Mechanism analysis of flow instability in multi-channel systems*

To investigate the underlying mechanism of flow oscillations, Figure 8 shows the outlet temperature (T_{out}) variation over time for each channel. Due to the initial maldistribution of mass flow, the outlet temperature of the low-flow channel C3 reaches as high as 800 K in the early stage, accompanied by small-amplitude temperature oscillations. In contrast, the outlet temperatures of C1 and C2 start relatively low and increase gradually, followed by rapid rises during t =12.5–17.5 s for C2 and t =17.5–22.5 s for C1. Once the outlet temperature of each channel reaches approximately 800 K, oscillatory behavior is triggered. This indicates that flow instability is initiated when the fluid crosses the pseudo-critical temperature, where thermophysical properties vary drastically along the flow path. Therefore, the sharp property gradients are identified as the key driver of flow instability.

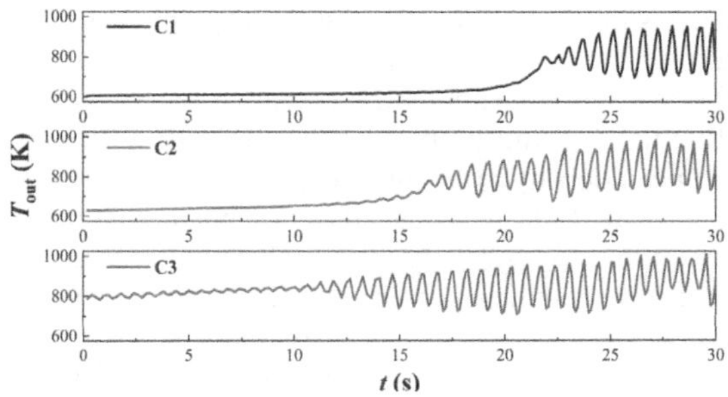

Figure 8. Time evolution of outlet temperatures in each channel.

Figure 9 illustrates the axial distributions of RP-3 fuel pyrolysis conversion rate (Z_{RP-3}) and fluid density (ρ) at different times. Influenced by temperature variation, the degree of fuel pyrolysis changes significantly over time. At t =10 s, the pyrolysis reaction primarily occurs in the low-flow channel C3. At t =17.5 s and t =22.5 s, significant pyrolysis is also observed in C2 and C1, respectively. The production of light hydrocarbon species further intensifies the variation in fluid density. As shown in Figure 9(b), under the combined effects of pyrolysis and pseudo-critical property transitions, the location where significant density drops occur shifts upstream for C1 and C2, and the rate of density decrease becomes more pronounced. Before reaching the outlet, the fluid density falls well below the pseudo-critical threshold, exhibiting gas-like characteristics.

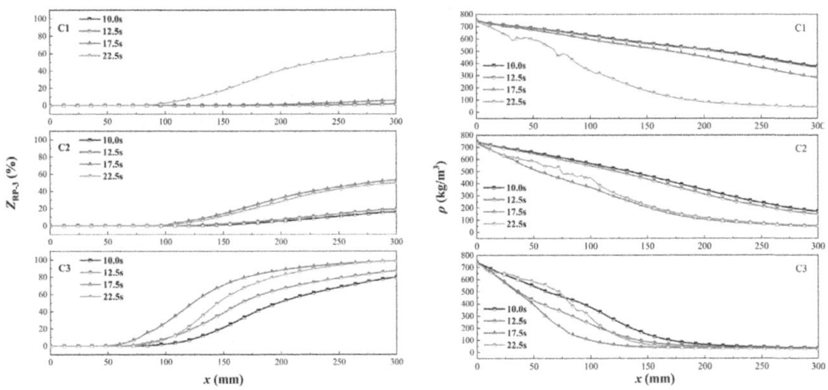

(a) Pyrolysis conversion rate (b) Density

Figure 9. Axial distributions of RP-3 fuel conversion rate and fluid density at various time instances.

Pressure drop is a key indicator of flow resistance in a channel. In regenerative cooling systems, total pressure drop (ΔP) consists mainly of frictional pressure drop (ΔP_{fr}) and acceleration pressure drop (ΔP_{ac}), the latter of which can be expressed as:

$$\Delta P_{\text{ac}} = \frac{m^2}{A^2}\left(\frac{1}{\rho_{\text{out}}} - \frac{1}{\rho_{\text{in}}}\right) \quad (6)$$

where m is the mass flow rate, A is the channel cross-sectional area, and ρ_{in}, ρ_{out} are the inlet and outlet fluid densities, respectively. This equation highlights that significant density gradients across the channel will result in large acceleration pressure drop. Additionally, strong density gradients can induce buoyancy-driven mixing, which increases frictional pressure drop. Since pseudo-critical fluid properties are highly sensitive to both temperature and pressure, even slight disturbances can modify the density profile, which in turn alters the acceleration and frictional pressure drops. These pressure drop changes are directly reflected in the channel flow resistance, which in turn affects the local mass flow rate. The resulting variations in flow rate modify the thermal environment, especially the temperature distribution, thereby amplifying property fluctuations and ultimately triggering flow oscillations.

In this study, pronounced density gradients were observed in all channels prior to the onset of flow oscillations. Therefore, large axial density gradients are a critical factor that makes the flow system highly susceptible to instability when subjected to external disturbances.

Figure 10 shows the time evolution of pressure drop (ΔP) in each channel. Between t = 0–10 s, all channels exhibit slight pressure fluctuations, but only C3 experiences flow instability, due to its initial large density gradient caused by the flow maldistribution. During t = 10–15 s, and t = 15–22.5 s, the pressure drops in C2 and C1 increase significantly, which corresponds to the earlier observed decline in flow rates. As flow rate decreases, fluid temperature rises, shifting the pseudo-critical zone upstream and accelerating fuel pyrolysis, resulting in steeper density gradients. As discussed previously, this further increases pressure drop and flow resistance, reinforcing the downward trend in flow rate. This forms a positive feedback loop among flow rate, density, and pressure drop: reduced flow leads to higher temperatures and greater density gradients, which increase resistance and further decrease flow rate, ultimately resulting in flow instability in that channel.

To further elucidate the inter-channel interactions in multi-channel systems, Figure 11 shows the turbulent kinetic energy (TKE) distributions along the channel cross-sections at different times. At t = 12.5 s, a high-TKE region appears at the outlet of channel C3, indicating intense turbulence and mixing, consistent with the onset of flow oscillations and secondary flows. A extensive TKE region is also observed in the downstream collector chamber, which begins to influence the downstream flow of adjacent channel C2. By t = 17.5 s, flow oscillations have developed in C2, with increased TKE at the outlet, and the

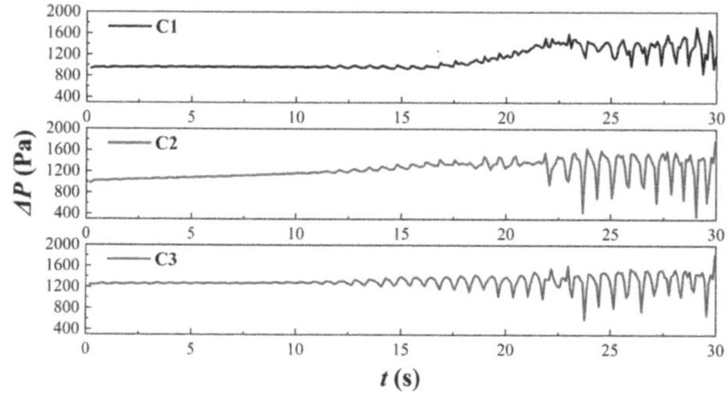

Figure 10. Time evolution of pressure drops in each channel.

high-TKE region expands further, now affecting channel C1. Eventually, at $t = 22.5$ s, strong turbulence is observed throughout the downstream region of all channels, indicating that the entire system has entered a state of sustained instability. This evolution confirms that flow instability in one channel can generate high-TKE regions in the downstream collector chamber, which significantly influence adjacent stable channels by increasing their downstream flow resistance. This resistance increase reduces flow rate, which then triggers the same positive feedback loop (flow rate–density–pressure drop) in the adjacent channels, ultimately causing widespread instability. Through this mechanism, instability initiated in one channel propagates to others, leading to system-wide self-excited oscillations.

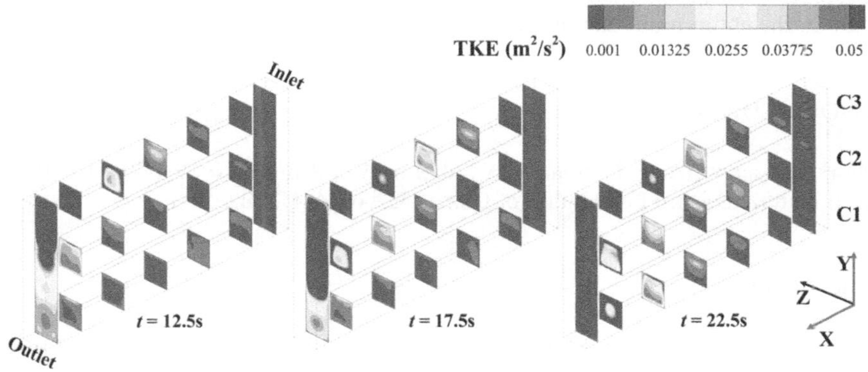

Figure 11. Streamwise distribution of turbulent kinetic energy at different times in each channel.

In summary, the flow oscillations observed in the multi-channel configuration are essentially triggered by initial instabilities in low-flow channels and subsequently propagate throughout the system. Gravity-induced maldistribution causes certain channels to operate near the pseudo-critical temperature region, where strong thermophysical property gradients make the flow highly sensitive to perturbations. Minor disturbances can lead to local density fluctuations, inducing acceleration and secondary flows that alter flow resistance and initiate small-scale oscillations. The resulting temperature rise further amplifies density gradients, intensifying the instability. As these unstable flows develop, they generate high turbulent kinetic energy regions in the downstream collector, which, in turn, increase the flow resistance of adjacent channels. This resistance increase reduces their flow rate, driving the channels into similar unstable conditions through a positive feedback loop involving flow rate, density, and pressure drop. As this process cascades through the system, the instability propagates across channels, eventually evolving into a self-excited, sustained oscillatory state driven by the coupled dynamics of pseudo-critical property variations and inter-channel interactions.

4 CONCLUSIONS

This study employed numerical simulations to investigate the evolution and underlying mechanisms of flow instability in a multi-channel system with supercritical RP-3 fuel. The main conclusions are summarized as follows:

(1) Under the influence of gravity, flow instability in the multi-channel configuration exhibits distinct evolutionary characteristics. The initial maldistribution of mass flow causes the low-flow channel to first develop oscillations, while high-flow channels initially

experience a gradual decline in flow rate followed by the onset of oscillations. The amplitude of oscillations is significantly larger in high-flow channels, whereas the dominant oscillation frequency remains nearly identical across all channels at approximately 1.48 Hz, indicating a typical density wave oscillation.

(2) Flow oscillations can propagate through the multi-channel system, eventually inducing global flow instability. Gravity-induced initial flow maldistribution leads to pronounced axial density gradients in the low-flow channels, making them highly susceptible to disturbances and prone to local instability. The resulting strong turbulent fluctuations in the downstream collector region significantly increase the flow resistance of adjacent, initially stable channels, thereby triggering a positive feedback loop among flow rate, fluid density, and pressure drop. This feedback mechanism drives the high-flow channels into instability in sequence, culminating in a globally self-excited and sustained oscillatory state in the entire multi-channel system.

ACKNOWLEDGMENT

This work was supported by National Key Laboratory of Ramjet (No. WDZC6142703202413); National Natural Science Foundation of China (52306064, U23B20108).

REFERENCES

Ambrosini, W. (2009). Discussion on the stability of heated channels with different fluids at supercritical pressures. *Nuclear Engineering and Design* 239(12): 2952–2963.

Du, W.D., Pan, Y., Wang, N., *et al.* (2025). Investigation on hydrodynamic characteristic and flow instability in regenerative cooling channels. *Applied Thermal Engineering* 269: 125983.

Fu, Y.C., Tao, Z., Xu, G.Q., *et al.* (2015). Experimental study of flow distribution for aviation kerosene in parallel helical tubes under supercritical pressure. *Applied Thermal Engineering* 90: 102–109.

Jiang, R.P., Liu, G.Z., Zhang, X.W. (2013). Thermal cracking of hydrocarbon aviation fuels in regenerative cooling microchannels. *Energy & Fuels* 27(5): 2563–2577.

Jiang, Y.G., Wang, Q., Zhang, D., *et al.* (2023). One-dimensional modelling and analysis of flow excursion of cracking hydrocarbon fuel in SCRamjet parallel cooling channels. *Fuel* 340: 127462.

Jing, T.T. (2018). Flow and thermal analyses of regenerative cooling of RBCC combustor based on hydrocarbon fuel. Xi'an: Northwestern Polytechnical University. (in Chinese)

Jin, Y.C., Wu, K., Fan, X.J. (2024). Flow instability of supercritical hydrocarbon fuel in a multi-channel cooling system. *Chinese Journal of Aeronautics* 37(2): 283–295.

Jin, Y.C., Wu, K., Lu, Y., *et al.* (2023). Investigation on the flow instability of supercritical hydrocarbon fuels in cooling channels. *Journal of Thermophysics and Heat Transfer* 37(3): 656–666.

Lin, D.S., Li, H.W., Cheng, Z.Y., *et al.* (2024). The heat transfer mechanism coupling pyrolysis and flow of hydrocarbon fuel in heated tube under supercritical pressure. *Applied Thermal Engineering* 247: 123017.

Wang, Z.B., Liu, X.W., Chen, J., *et al.* (2024). An overview of flow instability at supercritical pressure. *International Journal of Heat and Mass Transfer* 227: 125533.

Yan, J.J., Zhu, Y.H., Zhao, R., *et al.* (2018). Experimental investigation of the flow and heat transfer instabilities in n-decane at supercritical pressures in a vertical tube. *International Journal of Heat and Mass Transfer* 120: 987–996.

Zhou, W.X., Yu, B., Qin, J., *et al.* (2014). Mechanism and influencing factors analysis of flowing instability of supercritical endothermic hydrocarbon fuel within a small-scale channel. *Applied Thermal Engineering* 71 (1): 34–42.

Zhu, J.Q., Tao, Z., Deng, H.W., *et al.* (2015). Numerical investigation of heat transfer characteristics and flow resistance of kerosene RP-3 under supercritical pressure. *International Journal of Heat and Mass Transfer* 91: 330–341.

Zhu, Y.H., Peng, W., Xu, R.N., *et al.* (2018). Review on active thermal protection and its heat transfer for airbreathing hypersonic vehicles. *Chinese Journal of Aeronautics* 31(10): 1929–1953.

Fluid Mechanics and Industrial Applications – Fang (Ed.)
© 2026 The Author, ISBN: 978-1-041-15371-9

Application of CFD in the study of support interference of transport plane standard model

Liang Liu, Hongyang Chen, Peihong Zhang and Hui Zhao
China Aerodynamics Research and Development Centar, Mianyang, Sichaun, China

ABSTRACT: Support interference correction is always a key and difficult point in wind tunnel test. Based on the CFD software NNW-FlowStar of the National Numerical Wind Tunnel Project, this paper studies the influence of different support devices on the standard model of transport plane by means of numerical simulation, and analyzes the influence of different support devices on the aerodynamic characteristics and pressure distribution of transport plane, and analyzes the influence of blade thickness and leading edge sweep angle in the support devices on the aerodynamic force of the test model, which provides reference for the selection of support devices and data correction in subsequent tests.

1 INTRODUCTION

In the wind tunnel test, the model is supported by struts to measure the aerodynamic force in the wind tunnel test section. The existence of struts changes the flow field around the model, which makes the measured test data different from those without support devices. Therefore, it is always an important and difficult point in wind tunnel test to study the influence of support device interference and correct the test data. For a long time, researchers at home and abroad have carried out a large number of wind tunnel tests and numerical simulation studies on the interference correction of support devices in wind tunnel tests (Beyers, M E, 1985) (Ericssion, L E, 1982) (Kohzai, M, 2013) (Mouton, S, 2007) (Rist, M J, 1994), developed a variety of research methods on support interference, and applied them to engineering practice, accumulating rich valuable experience.

Based on the CFD software NNW-FlowStar of the National Numerical Wind Tunnel Project, this paper studies the influence of different support devices on the standard model of transport aircraft by means of numerical simulation, analyzes the influence of different support devices on the aerodynamic characteristics and pressure distribution of transport aircraft, and analyzes the influence of blade thickness and leading edge sweep angle in the support devices on the aerodynamic force of the test model, which provides reference for the selection of support devices and data correction in subsequent tests.

2 NUMERICAL METHODS

The NNW-FlowStar, a subtransonic supersonic flow solver developed by China Aerodynamics Research and Development Center (CARDC), is a large-scale parallel flow solver based on grid-centered unstructured hybrid grids and Reynolds-averaged Navier-Stokes equations. It can use mesh cells with arbitrary shapes and has great flexibility. The finite volume method is used to discretize the space, and the unknown variables are located at the center of the grid element. The implicit LU-SGS method or explicit Runge-Kutta

DOI: 10.1201/9781003679561-2

method is used to solve the discrete equations, and the FAS fusion multigrid method is used to accelerate the convergence. The solver has various options, such as discrete schemes of spatial convection and diffusion terms, time iterative methods, turbulence models and so on. At the same time, the software includes many different functional modules, which can be applied to aerodynamic calculation and type selection calculation of various aircraft.

Integral of N-S equation in three-dimensional Cartesian coordinate conservation form:

$$\frac{\partial}{\partial t}\int_{\Omega}\vec{W}d\Omega+\oint_{\partial\Omega}(\vec{F}_c-\vec{F}_v)dS=0 \tag{1}$$

Where, Ω denotes the volume of control volume, $\partial\Omega$ denotes the area of control volume closed surface, W denotes the conservation, $\vec{F}_c$ denotes the inviscid flux, $\vec{F}_v$ denotes the viscous flux.

The conserved variable is

$$\vec{W}=\begin{bmatrix}\rho\\ \rho u\\ \rho v\\ \rho w\\ \rho E\end{bmatrix} \tag{2}$$

The inviscid flux is

$$\vec{F}_c=\begin{bmatrix}\rho V\\ \rho uV+n_x p\\ \rho vV+n_y p\\ \rho wV+n_z p\\ \rho HV\end{bmatrix} \tag{3}$$

Where, $E=e+\frac{1}{2}(u^2+v^2+w^2)$, $H=E+\frac{p}{\rho}$

Viscous flux is

$$\vec{F}_v=\begin{bmatrix}0\\ n_x\tau_{xx}+n_y\tau_{xy}+n_z\tau_{xz}\\ n_x\tau_{yx}+n_y\tau_{yy}+n_z\tau_{yz}\\ n_x\tau_{zx}+n_y\tau_{zy}+n_z\tau_{zz}\\ n_x\Theta_x+n_y\Theta_y+n_z\Theta_z\end{bmatrix} \tag{4}$$

Roe flux difference splitting scheme is used for convection term, central scheme is used for viscous term, and implicit LU-SGS method is used for time discretization. Using MPI-based large-scale parallel computing technology and multi-grid method, the efficiency of flow field calculation is improved in many ways. SST two-equation model is used in turbulence model. SST model takes into account the influence of turbulent shear stress transport, and has a good ability to simulate strong reverse pressure gradient flow.

3 COMPUTATION MODEL AND MESHES

CAE-AVM standard model, designed by Chinese Aeronautical Establishment, is used to simulate the disturbance of strut system, with advanced regional aircraft and high-speed long-distance business aircraft as the background, and the layout is wing-body combination + nacelle + horizontal tail + vertical tail configuration (Figure 1), with a total length of 35.02 m, an extension of 30.183 m, a wing reference area of 100.454 m^2, an average aerodynamic chord length of 4.272 m and an aspect ratio of 9.0. The high aspect ratio supercritical wing is adopted, the leading edge sweep angle is 35, and the cruising Mach number is 0.85.

According to the calculation requirements, the engine of the aircraft is in a no intake and exhaust state, the surface grid of the aircraft is shown in Figure 2, the total grid of the whole aircraft is 16 million, and the grid near the aircraft head and engine is shown in Figure 3. In order to ensure the reliability of the calculation results, triangular prism and hexahedral meshes are generated body-fitted in the boundary layer region, tetrahedral meshes are used in the far field, and pyramid or tetrahedral meshes are used in the middle transition part. The prism layer structure can be clearly seen in the boundary layer. At the same time, anisotropic

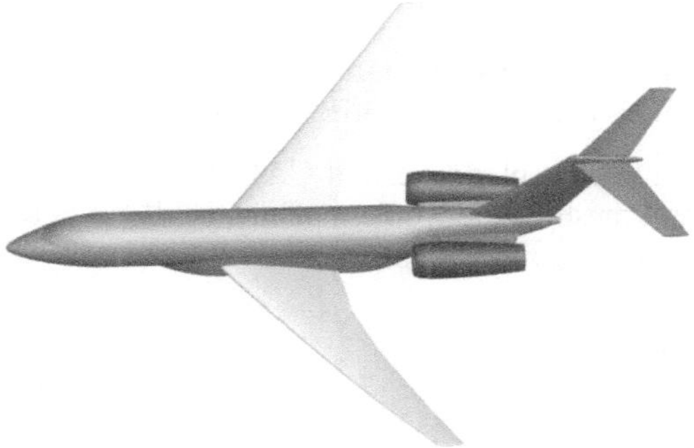

Figure 1. CAE-AVM standard model configuration.

Figure 2. CAE-AVM standard model global mesh (Full meshes 1600w).

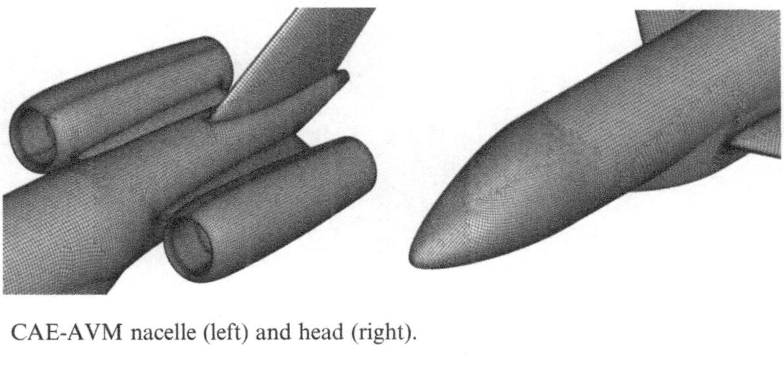

Figure 3. CAE-AVM nacelle (left) and head (right).

mesh technology is adopted in the head and local small-scale areas to realize the single-direction encryption of unstructured hybrid mesh in local flow sensitive areas. The wall-to-wall distance of the first layer of boundary layer is 3×10^{-7}m, corresponding to y+ = 1, the growth rate is 1.2, and the number of half-modular grid elements is 8356146, including tetrahedron 4317796, triangular prism 133306 and pyramid 382371. The Mach number of incoming flow is 0.85, the angle of attack is $-4°\sim8°$, and the calculated Reynolds number is 4.7×10^6. Aerodynamic reference values are: longitudinal reference length 0.24471 m, spanwise reference length 1.68903 m, reference area 0.31457 m^2, moment reference point (0, 0, 0).

In order to facilitate the selection of supporting rods in subsequent tests, two supporting methods commonly used in experiments are selected for calculation. The models of the above two supporting devices are shown in Figure 4 and Figure 5, The two support devices are described in detail as follows:

(1) CAE-AVM model with special Z support + direct head (ZG01);
(2) CAE-AVM model with belly braced blades and belly braced struts (ZG02).

Figure 6 shows the symmetric plane mesh of CAE-AVM model with special Z-support + direct head, and the total mesh of the whole machine is about 24.8 million. Figure 7 shows the enlarged local mesh of Z-support + direct head.

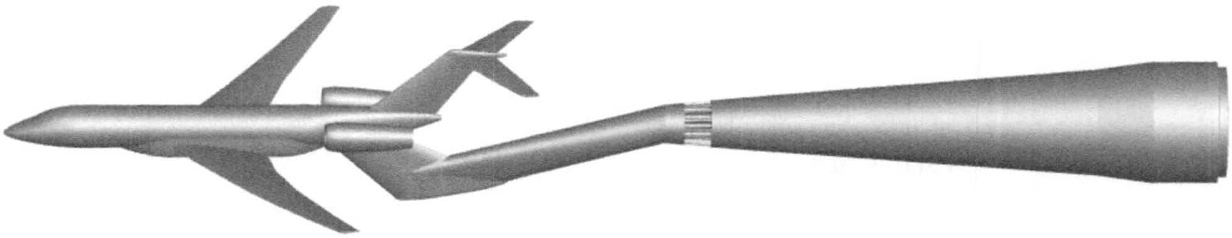

Figure 4. CAE-AVM model with special Z support + direct head (ZG01).

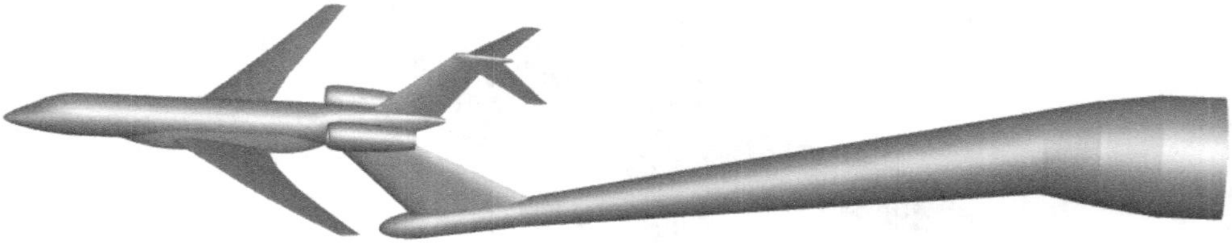

Figure 5. CAE-AVM model with belly braced blade + belly braced strut (ZG02).

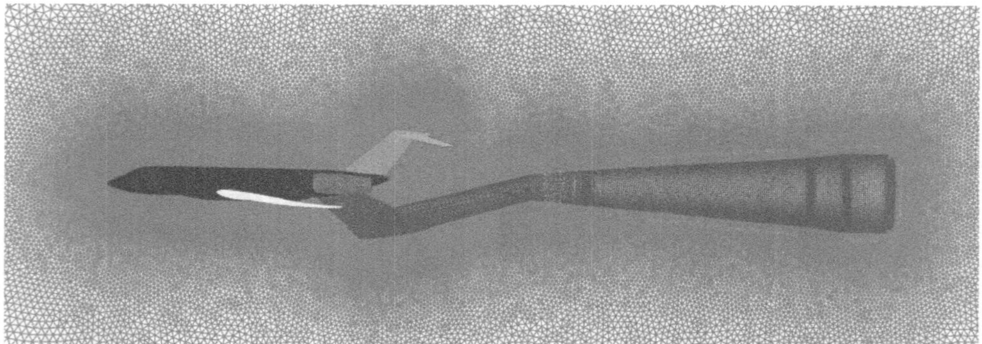

Figure 6. CAE-AVM model with special Z-support + direct head symmetric plane mesh.

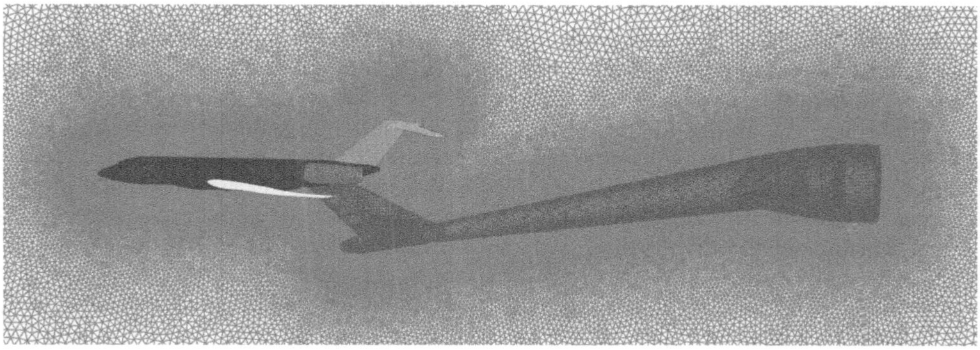

Figure 7. Symmetry plane mesh of CAE-AVM model with belly braced blades and belly braced struts.

4 ANALYSIS OF SUPPORT INTERFERENCE INFLUENCE

The comparison between the aerodynamic calculation results of the whole aircraft under different support devices and the results without support is shown in Figure 8. By comparison, it can be seen that the lift coefficient and drag coefficient calculated by ZG01 and ZG02 support methods have little difference, and they basically coincide; As a whole, there is little difference in lift between bracing and unbracing. Between -4°∼2°, the

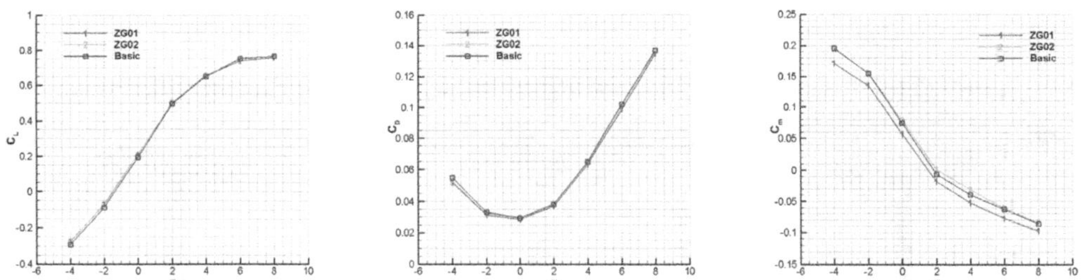

Figure 8. Influence of different support devices on aerodynamic force of transport aircraft.

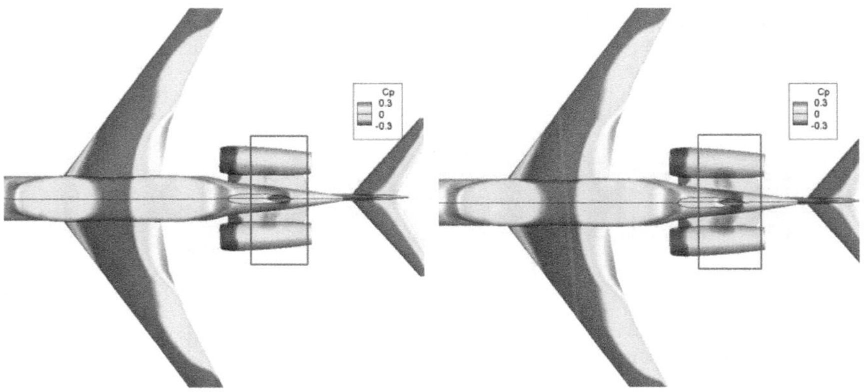

Figure 9. Comparison of pressure distribution on lower surface of ZG01(left) and ZG02(right).

lift coefficient of unbracing configuration is slightly less than that of bracing configuration, and when 2°~8°, the lift coefficient of unbracing configuration is slightly larger than that of bracing configuration. From the drag coefficient comparison results, the drag coefficient of unsupported configuration is slightly greater than that of supported configuration; According to the torque curve, the pitching moment coefficient of the whole aircraft under the two support modes is quite different, mainly because different support modes will affect the pressure distribution on the lower surface of the aircraft (Figure 9), and then influence the torque. The torque results of ZG02 are closer to those of the unsupported configuration, with only certain differences at 2° and 4°, and the results basically coincide at other angles of attack. Based on the aerodynamic comparison results, ZG02 support device can be preferentially selected for testing in subsequent wind tunnels, which has minimal interference to the aerodynamic force of the whole aircraft.

On the basis of ZG02 support device, the influence of ventral support blade thickness and blade leading edge sweep angle on wind tunnel interference is analyzed. As shown in Figure 10, there are three thicknesses for blades. From the comparison curve in Figure 11, the normal force of the whole machine under different blade thicknesses is basically the sa me as that without support device, with little difference; There is a certain difference in axial force, and the axial force of the whole machine decreases slightly after installing blades; On the whole, when the blade thickness is 15%, the axial force difference between the whole machine and that without supporting device is the smallest. Figure 12 shows the spatial streamline distribution at the support area when the thickness of the ventral braced blades is 10%, 13% and 15%. The blue streamline area is the flow separation area. It can be seen that the greater the blade thickness, the larger the separation area.

Figure 10. Different thickness of leaves (left 10%, middle 13%, right 15%).

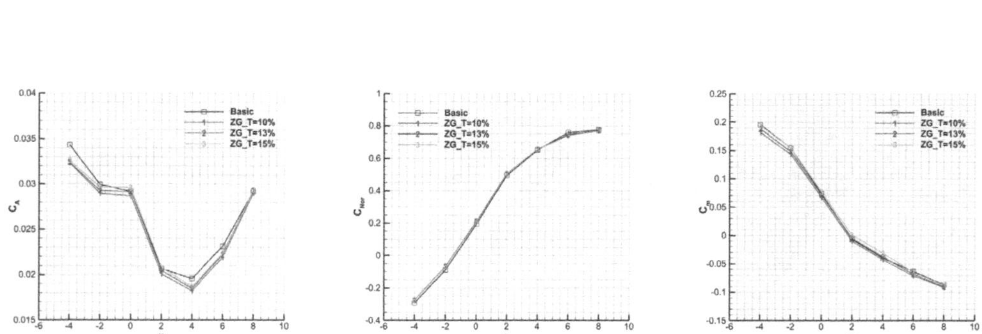

Figure 11. Influence of different blade thickness on aerodynamic force.

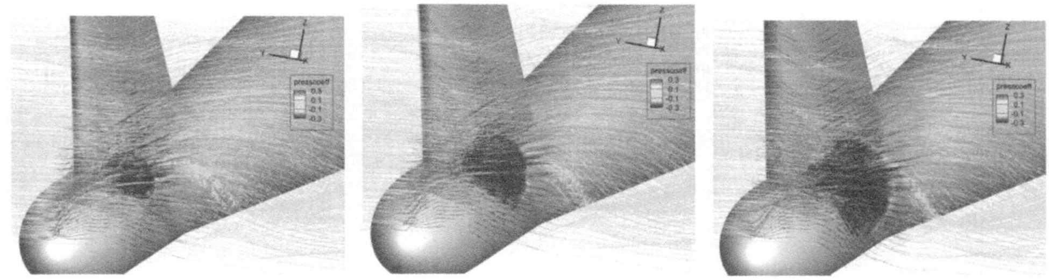

Figure 12. Streamline distribution of supporting space for different blade thickness (left 10%, middle 13%, right 15%).

Besides, blade leading edge sweep angle on wind tunnel interference is also analyzed. As shown in Figure 13, there are three sweep angles of leading edge. From the comparison curve in Figure 14, the normal force and moment of the whole machine under different blade leading edge sweep angles are basically the same as those without supporting devices, with little difference; As a whole, when the sweep angle of blade leading edge is 35°, the axial force difference between the whole machine and the whole machine without supporting device is the smallest. As the wing and fuselage contribute the most to the axial force of the whole aircraft, it can be seen from Figure 15 that the axial forces of wing components with different sweep angles are almost coincident, while the axial forces of fuselage are different, and the axial forces of fuselage without support are less different when the sweep angle is 35°. Figure 16 shows the pressure distribution on the lower surface of the aircraft with blade leading edge sweep angles of 35°, 47.42° and 60°, and unsupported configuration at an angle of attack of -2°. It can be seen that the fuselage pressure distribution is different. Figure 17 shows the streamline distribution of the supporting surface at the blade leading edge sweep angles of 35°, 47.42° and 60°. It can be seen that when the blade leading edge sweep angle is 35°, there is no flow separation near the leading edge, but the other two sweep angles have flow separation.

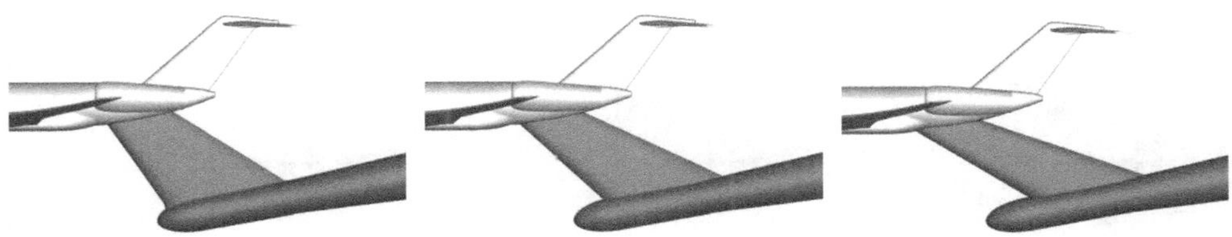

Figure 13. Different sweep angles of leading edge (left 35°, middle 47.42°, right 60°).

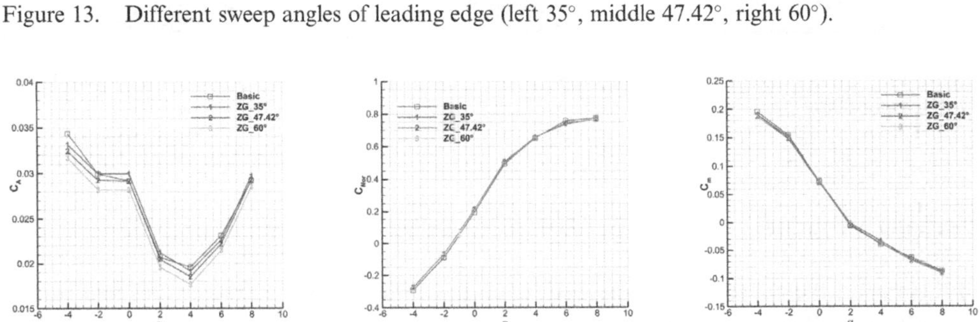

Figure 14. Influence of different sweep angles of leading edge on aerodynamic force.

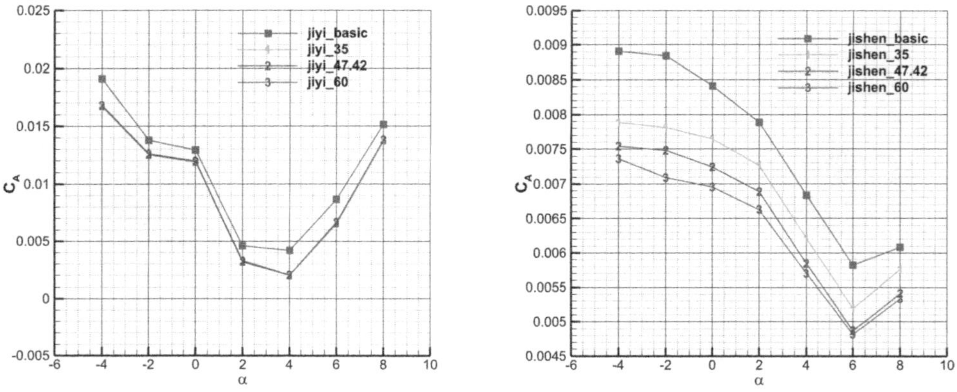

Figure 15. Comparison of axial forces of wing (left) and fuselage (right).

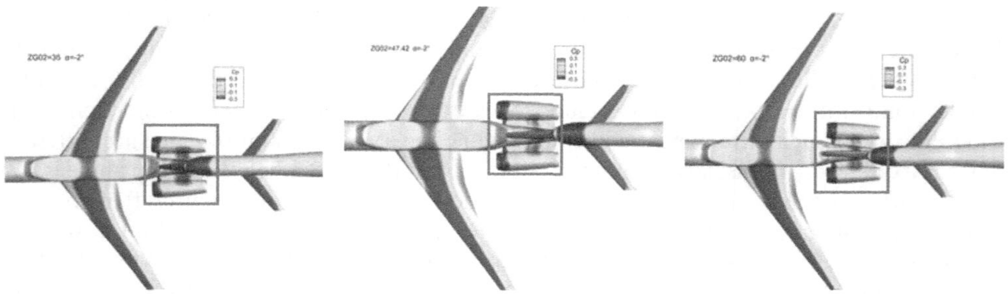

Figure 16. Pressure distribution on lower surface of aircraft with different sweep angle (left 35°, middle 47.42°, right 60°)

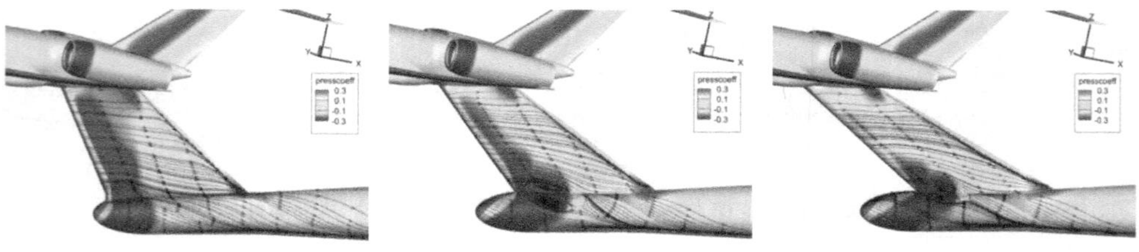

Figure 17. Streamline distribution of support surface with different sweep angle(left 35°, middle 47.42°, right 60°).

5 CONCLUSION

In this paper, the effects of support devices on the aerodynamic characteristics and surface pressures of a typical conveyor with support devices are studied, and the effects of blade thickness and leading edge sweep angle in the support devices on the aerodynamic force of the test model are analyzed. Compared with Z-type support, abdominal support has the least aerodynamic interference to transport aircraft; When the thickness of the vane supported by the abdomen is 15%, the axial force interference to the conveyor is the smallest; When the sweep angle of the leading edge of the blade is 35°, the interference to the axial force of the transport aircraft is minimal.

REFERENCES

Beyers, M E. (1985). *Some Recent NAE Experiences of Support Interference in Dynamic Tests.* NAE Report LTR-UA-83, Ottawa, Ontario.

Ericssion, L E. Reding, J P. (1982). *Review of support interference in dynamic tests.* 12th Aerodynamic Testing Conference.

Kohzai, M. Ueno, M. Koga, S. (2013). *Wall and support interference corrections of NASA common research model wind tunnel tests in JAXA.* 51st AIAA Aerospace Science Meeting Including the New Horizons Forum and Aerospace Exposition.

Mouton, S. (2007). *Numerical investigation of model support interference in subsonic and transonic wind tunnels.* 8th ONERA-DLR Symposium ODAS.

Rist, M J. (1994). *Computational investigation of wind tunnel support interference.* 25th Plasmadynamics and Lasers Conference.

Fluid Mechanics and Industrial Applications – Fang (Ed.)
© 2026 The Author, ISBN: 978-1-041-15371-9

Photocatalytic fluid chemical mechanical polishing achieves ultra-smooth surfaces for additively manufactured AlSi10Mg

Zeyun Wang, Zhenyu Zhang and Jianan Xu
College of Mechanical and Electrical Engineering, Harbin Engineering University, Harbin, China.

ABSTRACT: This study investigates the chemical mechanical polishing (CMP) behavior of AlSi10Mg alloys using a series of nine designed slurries. The experimental results reveal that the surface quality is strongly dependent on the synergistic action of abrasives, oxidants, and organic additives. Surface roughness (Sa) measurements indicate that the optimized slurry, containing silica and ceria abrasives, polyethylene glycol, citric acid, hydrogen peroxide, and serine under light-assisted catalysis, achieves the smoothest surface with an Sa of 0.242 nm. Electrochemical analyses demonstrate that the addition of oxidants and amino acids regulates both corrosion potential and current density, thereby balancing passivation and material removal. FTIR spectra further confirm the involvement of serine functional groups and the formation of Al-O, Si-O, and Si-O-Ce bonds, which facilitate surface complexation and chemical modification. A comprehensive CMP mechanism is proposed in which serine chelation, peroxide oxidation, and light-enhanced ceria catalysis synergistically promote efficient chemical reactions, while silica abrasives ensure mechanical removal. These findings establish an effective strategy for achieving near-atomic-level planarization of lightweight AlSi10Mg alloys.

1 INTRODUCTION

AlSi10Mg has become one of the most widely used aluminum alloys in additive manufacturing due to its excellent specific strength, lightweight properties, and good thermal stability, demonstrating significant potential particularly in aerospace, optical components, and precision devices (Q. Gu, 2024; Z. Wang, 2025). However, during selective laser melting (SLM) processing, this material forms complex dendritic structures and uneven silicon phase distribution. These defects make it difficult to directly meet the requirements for ultra-smooth surface applications. Traditional methods can improve surface quality but often suffer from low material removal rates, prone to surface waviness, or introduce subsurface damage, limiting their application in high-precision optical processing (Z. Xue, 2023). In contrast, chemical mechanical polishing (CMP) achieves atomic-level surface leveling through the synergistic action of chemical reactions and mechanical forces, making it a potential key technology for enhancing the surface quality of AlSi10Mg additive manufacturing parts (Lei, X, 2025).

Existing research indicates that CMP demonstrates certain feasibility in surface finishing of aluminum alloys and other metallic materials, with its core mechanism residing in the synergistic interaction between chemical reagents and abrasives within the slurry (Z. Wang, 2025). Chemical components soften the surface through oxidation or complexation, while abrasive particles achieve material removal via micro-cutting (T. Li, 2021). However, significant challenges persist when applying CMP to AlSi10Mg alloys. The multiphase microstructure of Al-Si-Mg alloys and the limitations of conventional slurries together lead

DOI: 10.1201/9781003679561-3

to uneven removal, selective corrosion, unstable oxide films, and surface defects (Z. Wu, 2023). Consequently, conventional CMP systems struggle to simultaneously achieve high efficiency and high surface flatness, necessitating the development of new technical approaches.

In recent years, photocatalytic technology has been introduced into CMP systems to enhance chemical reactivity and surface removal efficiency (C. Lin, 2025; H. Li, 2025). Unlike approaches that merely incorporate photocatalysts into the slurry, this study employs cerium oxide (CeO_2) as an abrasive that combines both mechanical and photocatalytic functions. Under the combined effects of light irradiation and hydrogen peroxide, CeO_2 not only promotes the generation of active species such as ·OH through the Ce^{4+}/Ce^{3+} redox cycle, thereby accelerating surface reactions with Al and Si phases, but also functions as an abrasive particle to achieve stable mechanical removal (X. Cui, 2023). A novel photocatalytic fluid for chemical-mechanical polishing slurry has been developed, comprising silicon dioxide and cerium oxide abrasives, hydrogen peroxide, citric acid, polyethylene glycol, and glycine. The novel slurry has achieved an extremely smooth surface for the AlSi10Mg alloy, and the mechanism of its action and its performance advantages have been studied.

2 EXPERIMENTAL DETAIL

Silica and ceria abrasives, hydrogen peroxide, citric acid, polyethylene glycol, and glycine were procured from Shanghai Sinopharm Co., Ltd. The average particle size of all abrasives was 50 nm. The polished specimens were additive-manufactured AlSi10Mg alloy. The polishing specimens were rectangular blocks measuring $10 \times 10 \times 5$ mm, affixed to aluminum alloy specimen holders using paraffin wax. Prior to polishing, the abrasive, polyethylene glycol, and deionized water were mixed and subjected to 15 minutes of ultrasonic dispersion. Hydrogen peroxide, citric acid, and glycine were then added, followed by continuous stirring of the mixture on a magnetic stirrer. To investigate the effects of different slurry compositions on polishing quality, nine polishing experiments were conducted as shown in Table 1.

Table 1. Composition of CMP slurry.

Group	Slurry composition	pH
S1	SiO_2	7
S2	CeO_2	7
S3	SiO_2+ CeO_2	7
S4	SiO_2+ CeO_2+PEG	7
S5	SiO_2+ CeO_2+PEG + citric acid	5
S6	SiO_2+ CeO_2+PEG + citric acid +H_2O_2	5
S7	SiO_2+ CeO_2+PEG + citric acid + glycine	5
S8	SiO_2+ CeO_2+PEG + citric acid +H_2O_2+ glycine	5
S9	SiO_2+ CeO_2+PEG + citric acid +H_2O_2+ glycine + photocatalysis	5

Polishing experiments were conducted using a precision grinding and polishing machine manufactured by Shenyang Kejing Automation Equipment Co., Ltd. Polishing parameters included a pressure of 30 kPa, a rotational speed of 85 rpm, a slurry flow rate of 15 mL/min, and a polishing duration of 20 min. Following polishing, sample surfaces were cleaned with ethanol. Surface roughness (Sa) was characterized using a non-contact optical profilometer (SuperView W1 series, CHOTEST, China). The Tafel polarization curve of the slurry was measured using a CHI760E electrochemical workstation (Shanghai Chenhua Instruments Co., Ltd.). Fourier transform infrared spectroscopy (FTIR; Nicolet iS20, Thermo Scientific, USA) was used to investigate the changes in surface functional groups of AlSi10Mg.

3 RESULTS AND DISCUSSION

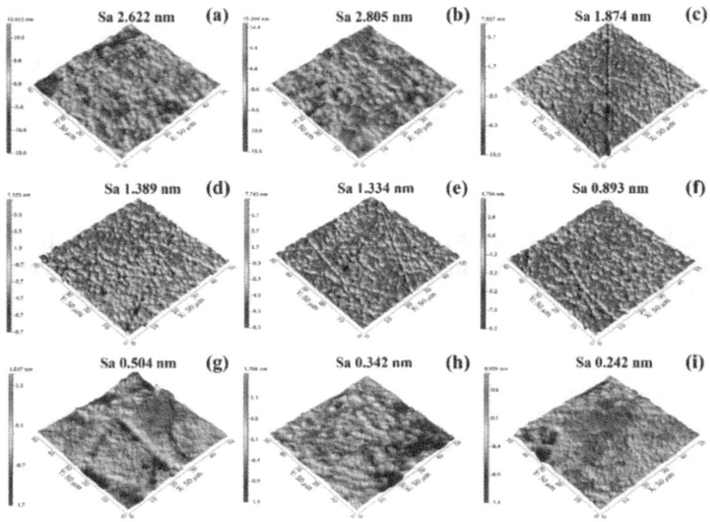

Figure 1. Surface morphology after polishing with different slurries.

The surface morphologies of the Al-Si-Mg samples after CMP with nine different slurry formulations are shown in Figure 1(a–i), corresponding to S1-S9. Clear differences in Sa can be observed as the slurry composition becomes progressively more complex. When polished with single-abrasive slurries (Figure 1(a, b)), the surfaces exhibited relatively high roughness due to the absence of effective chemical modification. The mixed-abrasive slurry (Figure 1 (c)) achieved slightly smoother surfaces, indicating that the synergistic effect between SiO_2 and CeO_2 contributes to more effective mechanical action. The introduction of polyethylene glycol (Figure 1(d)) noticeably reduced the roughness, which can be attributed to steric stabilization that improves abrasive dispersion. With the addition of citric acid (Figure 1(e)), the surface became significantly smoother, demonstrating that citric acid effectively chelates Al ions and suppresses uncontrolled corrosion.

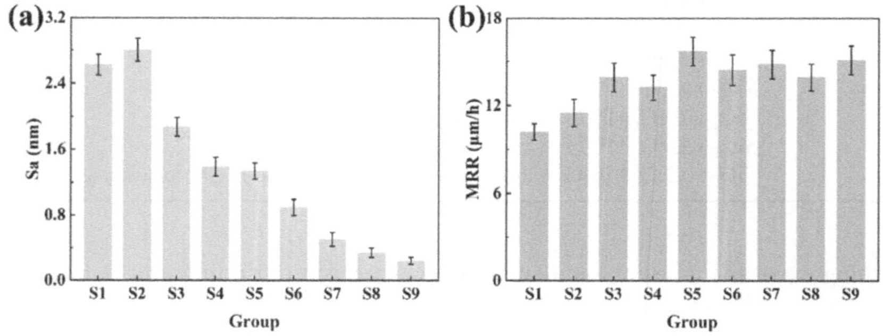

Figure 2. Sa and MRR values after polishing with different slurries.

The inclusion of H_2O_2 (Figure 1(f)) further decreased the roughness, as the oxidant promoted the formation of a thin, soft oxide layer that is easily removed by abrasives. Glycine (Figure 1(g)) provided an additional polishing effect by forming amino-metal complexes, which facilitated uniform dissolution of asperities. When H_2O_2 and glycine were combined (Figure 1(h)), the slurry achieved one of the lowest roughness values, suggesting a strong synergistic interaction between oxidation and complexation. Finally, under light-assisted conditions (Figure 1(i)), the surface exhibited the smoothest morphology among all groups. This can be explained by the photo-catalytic activation of CeO_2 and the photolysis of H_2O_2, which together enhance the generation of reactive oxygen species and accelerate chemical reactions at the surface. These results highlight that the optimized slurry should integrate both chemical additives and mechanical components. The progressive reduction in Sa from S1 to S9 confirms the effectiveness of a multi-component synergistic approach in achieving ultra-smooth Al-Si-Mg surfaces.

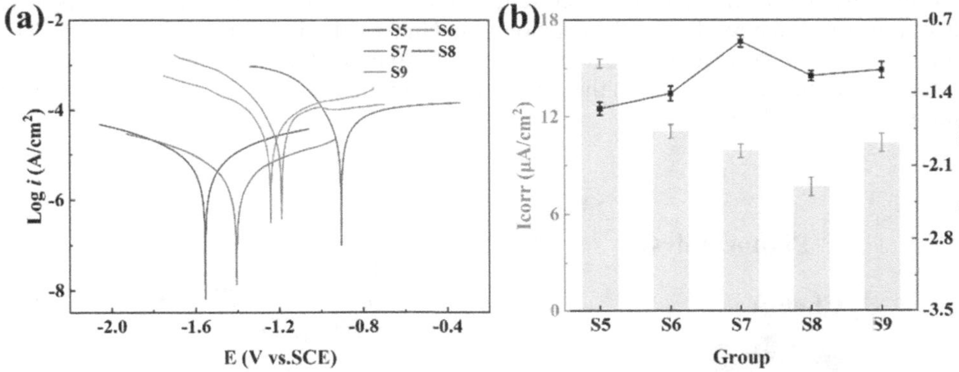

Figure 3. Electrochemical testing after immersion in S5-S9 slurries. (a) Tafel curves, (b) Corrosion current and corrosion voltage.

As shown in Figure 2, the performance of the nine slurries exhibits clear variations during the CMP of the Al-Si-Mg alloy. In Figure 2(a), the Sa first decreases and then slightly increases, with the lowest values obtained for slurries S5-S7. This indicates that the synergistic effect of additives such as polyethylene glycol, citric acid, and hydrogen peroxide effectively suppresses scratches and achieves a smoother surface. Correspondingly, Figure 2 (b) shows that the material removal rate (MRR) is relatively low for S1-S3, but reaches its maximum for S6-S8, suggesting that the proper introduction of oxidants and complexing agents significantly accelerates the removal process. In contrast, S9 shows a distinct behavior, as the introduction of light irradiation catalyzes the CeO_2 abrasives to enhance the

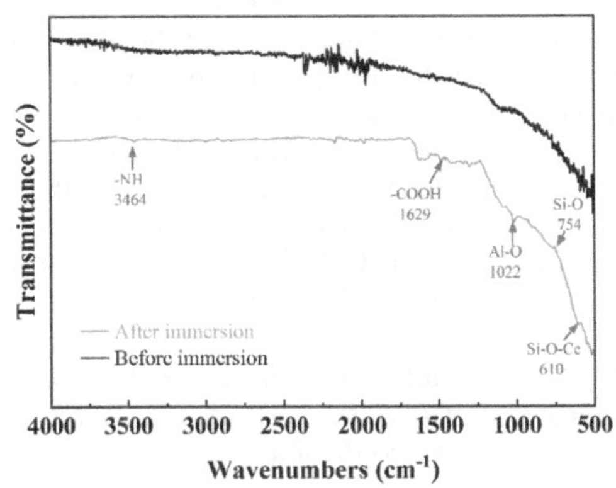

Figure 4. FTIR curve after S9 slurry immersion.

oxidation and polishing of Si. This photocatalytic effect leads to a differentiated removal mechanism compared with the other slurry systems, highlighting the potential of light-assisted CMP.

As shown in Figure 3, the electrochemical behavior of the Al-Si-Mg alloy surface in slurries S5-S9 was evaluated by Tafel polarization curves and the corresponding corrosion parameters. In Figure 3(a), the Tafel curves exhibit noticeable shifts in both anodic and cathodic branches with the addition of different additives, indicating that the electrochemical reactions at the surface are strongly dependent on slurry chemistry. As summarized in Figure 3(b), the corrosion current density initially increases from S5 to S7, suggesting that the incorporation of oxidants and organic additives promotes interfacial reactions and facilitates the material removal process. When hydrogen peroxide and glycine are simultaneously introduced, the corrosion current decreases slightly, implying a competitive adsorption effect that moderates the reaction rate. Under light-assisted conditions with CeO_2 abrasives, both the corrosion current and potential exhibit distinct changes, confirming that photocatalytic enhancement alters the surface oxidation mechanism of Si and contributes to a more efficient CMP process.

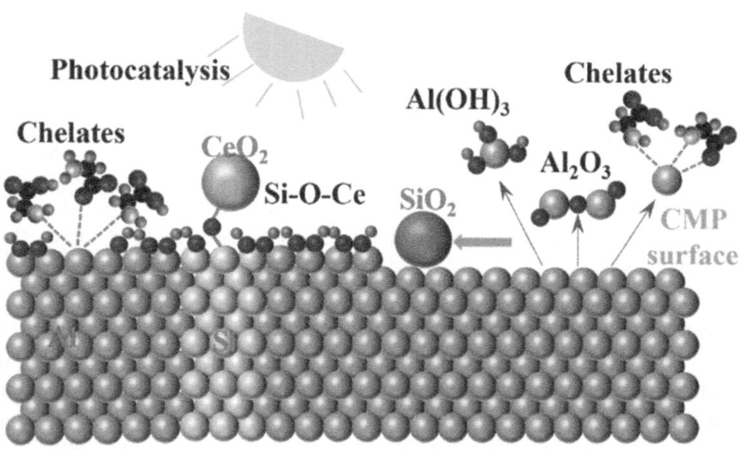

Figure 5. Schematic diagram of CMP mechanism.

As shown in Figure 4, the FTIR spectrum of the Al-Si-Mg alloy after immersion in the S9 slurry reveals characteristic absorption bands corresponding to both organic and inorganic functional groups. The peaks associated with serine molecules can be clearly identified, indicating that the amino acid adsorbs onto the alloy surface and participates in the surface reactions during CMP. In addition, the presence of Al-O and Si-O stretching vibrations suggests that both aluminum oxide and silicon oxide species are formed as a result of chemical modification of the alloy surface. More importantly, the appearance of the Si-O-Ce band confirms the interfacial interaction between the silicon species and CeO_2 abrasives, which is consistent with the proposed photocatalytic mechanism under light irradiation. This bonding indicates that CeO_2 not only acts as a mechanical abrasive but also chemically couples with oxidized silicon, thereby enhancing the synergistic effect between mechanical abrasion and chemical passivation. The mechanism of CMP is shown in Figure 5. Overall, the FTIR results corroborate the electrochemical findings, demonstrating that the combined action of serine adsorption, oxide formation, and Si-O-Ce interfacial bonding under irradiation contributes to the unique polishing behavior of the S9 slurry.

4 CONCLUSION

A systematic evaluation of nine slurry formulations demonstrated that the polishing performance of Al-Si-Mg alloys can be significantly enhanced through the rational integration of abrasives and chemical additives. The optimized slurry exhibited the lowest surface roughness (0.242 nm), highlighting the benefits of combining silica and ceria abrasives with hydrogen peroxide and serine under light-assisted conditions. Electrochemical testing confirmed that controlled corrosion and passivation were achieved by fine-tuning the slurry chemistry, while FTIR analysis revealed strong evidence of surface bonding involving serine and Si-O-Ce linkages. The proposed CMP mechanism underscores the cooperative roles of serine chelation, oxidative reactions, and catalytic activity of ceria abrasives, together with the mechanical contribution of silica particles. This work provides a sustainable and high-precision CMP pathway for Al-Si-Mg alloys, with potential implications for advanced electronic and aerospace manufacturing.

ACKNOWLEDGEMENTS

Zeyun Wang: Investigation, Formal analysis, Data curation, Visualization. Zhenyu Zhang: Funding acquisition, Project administration, Methodology, Supervision, Conceptualization. Jianan Xu: Formal analysis, Investigation.

REFERENCES

Cui X., Zhang Z., Yu S., Chen X., Shi C., Zhou H., Meng F., Yu J., Wen W. (2023). Unprecedented atomic surface of silicon induced by environmentally friendly chemical mechanical polishing.

Gu Q., Zhang Z., Zhou H., Yu J., Wang D., Feng J., Shi C., Yang J., Qi J. (2024). A novel approach of jet polishing for interior surface of small-grooved components using three developed setups, *Int. J. Extrem. Manuf.* 6 025101. https://doi.org/10.1088/2631-7990/ad1bba.

Lei, X., Zhang, Z., Zhou, H., Yang, X., Hu, P., He, L., ... & Pang, Z., Novel eco-friendly chemical mechanical polishing of aluminum alloy for achieving close atomic surface. *Wear*, 578 (2025) 206174.

Li H., Yang J., Lu J., Sun J., Hu D., Yan Q. (2025). Study on the process parameters of UV photocatalytic chemical mechanical polishing for single-crystal AlN, *Diamond and Related Materials* 156, 112407. https://doi.org/10.1016/j.diamond.2025.112407.

Li T., Sun H., Wang D., Huang J., Li D., Lei F., Sun D. (2021). High-performance chemical mechanical polishing slurry for aluminum alloy using hybrid abrasives of zirconium phosphate and alumina, *Applied Surface Science* 537 147859. https://doi.org/10.1016/j.apsusc.2020.147859.

Lin C., Xie J., Lu J., Yan Q. (2025). Study on the effect of process parameters and removal behavior of single-crystal SiC polishing based on a photocatalytic fixed polishing plate, *Diamond and Related Materials* 153, 111985. https://doi.org/10.1016/j.diamond.2025.111985.

Wang Z., Zhang Z., Hu P., Liu G., Hu J., Xu J., Cai H., Pang Z., Ding P. (2025). An atomic surface of an aluminum alloy induced by novel green chemical mechanical polishing using hybrid rare earth abrasives and mechanisms unraveled, *Nanoscale*. https://doi.org/10.1039/d5nr02599k.

Wang Z., Zhang Z., Zhou H., Han D., Shi C., Chen L., Yao J., Yu S., Xu J. (2025). Close atomic surface on aluminum alloy using green chemical mechanical polishing with synergistic effect between yttria and silica abrasives, *Applied Surface Science* 681 161586. https://doi.org/10.1016/j.apsusc.2024.161586.

Wu Z., Shen J., Wu X., Peng Y., Lin S., Huang L., Huang X., Zhu L. (2023). Effect of dispersion performance of polishing slurry on the polishing quality of glass-ceramics in bonnet polishing, *The International Journal of Advanced Manufacturing Technology* 127 (2023) 107–121. https://doi.org/10.1007/s00170-023-11532-9.

Xue Z., Lai M., Xu F., Fang F., (2024). Influence factors and prediction model of surface roughness in single-point diamond turning of polycrystalline soft metal, *Journal of Materials Processing Technology* 324 (2024) 118256. https://doi.org/10.1016/j.jmatprotec.2023.118256.

Fluid Mechanics and Industrial Applications – Fang (Ed.)
© 2026 The Author, ISBN: 978-1-041-15371-9

The influence of variable geometric lip on supersonic inlet performance

Bing Chen, Liang Liu, Xiaojun Wu, Wenhao Zhang and Hongyin Jia
Computational Aerodynamics Institute, China
Aerodynamics Research and Development Center, Mianyang, Sichuan, China
State Key Laboratory of Aerodynamics, Mianyang, Sichuàn, China

ABSTRACT: The inlet is the key component that affects the internal and external flow performance of air-breathing vehicle. The wide-envelope flight requires the inlet to have geometric adjustment ability, and the deformation parameters are very important to the internal flow performance. In this paper, the variable geometry inlet with deformed lip is taken as the object. Through numerical simulation, the influence of the main deformation parameters of the inlet lip on the inlet performance is studied, and the influence of amplitude and frequency on the inlet performance is obtained. The results show that the rotation frequency of the lip has a great influence on the inlet performance, which not only leads to the oscillation of the inlet performance, but also induces a larger oscillation of the inlet performance at low frequency. The relevant research results can be used to support and improve the design criteria of inlet with variable lip, which has certain engineering guiding significance.

1 INTRODUCTION

Inlet is an important component of air-breathing vehicle. The coupling between internal and external flow is serious for airframe-propulsion integrated configuration, and the internal flow performance is highly sensitive to the change of shape parameters. The structural vibration or geometric deformation of inlet will induce violent oscillation of internal flow performance, which will bring great negative impact on the design. Therefore, it is of great practical significance to study the influence of inlet geometric deformation on inlet performance.

The aeroelasticity induced deformation and vibration is an important challenge for the inlet due to complex mechanical and thermal effects during flight. With the in-depth study of integrated coupling mechanism, researchers gradually realize that the elastic deformation of the fuselage and aerothermal problems have a significant impact on the integration of internal and external flow For air-breathing vehicle. The effects of aeroelasticity and aerothermo-elasticity on internal and external flow coupling are studied. McNamara *et al.* (2008) studied the aeroelasticity and thermo-aeroelasticity in hypersonic flow, and Culler and McNamara (2010) studied the coupling problem of thermo-aeroelasticity among flow field, thermal field and structural field in hypersonic flow. Aiming at the coupling problem of aeroelasticity and internal and external flow integration of air-breathing aircraft, Clark *et al.* (2006) established an analysis model of aeroelasticity and integration, and studied the response characteristics of the integrated configuration under aeroelasticity by combining CFD and engineering method, and made clear the influence of fuselage elastic deformation on the integrated performance. Relevant tools can be used to analyze the dynamic characteristics of air-breathing vehicle of airframe-propulsion integrated configuration.

DOI: 10.1201/9781003679561-4

The influence of structural deformation on inflow performance is difficult to ignore because of the complex coupling. Kline, Palacios and Alonso (2014) studied the influence of inlet's deformation on the performance of scramjet. The results show that the deformation of 1.63 μ m will lead to 4.5% thrust change. However, the structural vibration of inlet's compression surface not only induces internal flow oscillation, but also has certain hysteresis effect. Long-term structural oscillation will lead to the increase or decrease of internal flow performance oscillation, which may eventually lead to disastrous consequences. Kun, Zheng-yin and Zhan (2017) studied the influence of amplitude, wavelength and frequency of panel on combustion efficiency and total pressure loss for scramjet, and further clarified the sensitivity of ramjet performance to structural vibration.

Variable geometry techniques are generally adopted for supersonic inlet operating in a wide speed range, including variable compression surface, variable lip and variable throat, etc. Among them, the lip layout will greatly affect inflow of inlet, so the lip adjustable technology is a typical variable geometry scheme. Therefore, the geometric deformation of the lip is an important factor that must be considered in the process of design, and its influence on the integration performance of internal and external flow is mainly manifested in the influence of different deformation modes.

The paper studies the influence of structural vibration of the lip on the inlet performance. The influence of amplitude and frequency on the inlet flow field is studied by numerical simulation. The oscillation characteristics of inlet performance under the influence of various factors are clarified.

2 PROBLEM DESCRIPTION

The complicated flow phenomena include shock-shock wave and shock-expansion wave for the inlet with variable geometry lip. The characteristics are shown as in Figure 1.

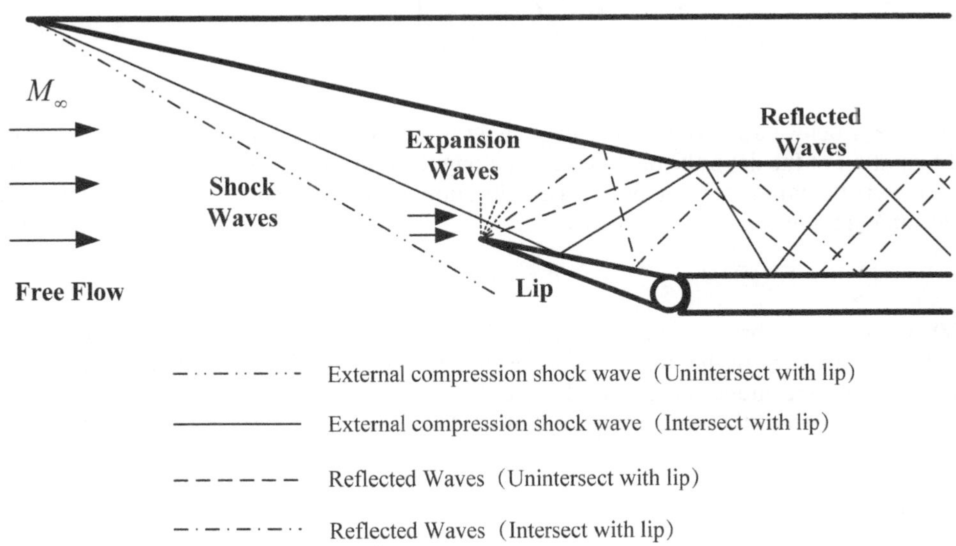

Figure 1. Fluid-structure coupling characteristics of inlet lip.

When the lip rotates around the axis, it will inevitably cause the disturbance of the internal flow, which will lead to the periodic or aperiodic oscillation of the inlet performance. Under the action of continuous oscillating airflow, it will lead to the thrust loss and specific impulse decline of the engine.

The influence factors of lip on inlet performance mainly include lip reflected shock angle and shock intensity, which is manifested in the rotation amplitude of the lip. In addition, the rotation of lip around the axis will produce serious unsteady phenomenon. It will lead to oscillation of internal flow performance, which will bring great difficulties to inlet starting. Therefore, it is necessary to study the influence of vibration elements of inlet lip on inlet performance. The motion elements mainly include amplitude and frequency when lip rotates.

3 METHODOLOGY

The flow characteristics of the inlet are complex, so the boundary layer effect must be considered in the numerical solution process. The aerodynamic analysis adopts three-dimensional NS equation considering viscosity. Its conservation form can be expressed as

$$\frac{\partial \mathbf{U}}{\partial t}+\frac{\partial \mathbf{E}}{\partial x}+\frac{\partial \mathbf{F}}{\partial y}+\frac{\partial \mathbf{G}}{\partial z}=\frac{1}{\mathrm{Re}}\left(\frac{\partial \mathbf{E}_v}{\partial x}+\frac{\partial \mathbf{F}_v}{\partial y}+\frac{\partial \mathbf{G}_v}{\partial z}\right) \tag{1}$$

Where,

$$\mathbf{U}=\begin{bmatrix}\rho\\ \rho u\\ \rho v\\ \rho w\\ \rho E\end{bmatrix},\ \mathbf{E}=\begin{bmatrix}\rho u\\ \rho u^2+P\\ \rho uv\\ \rho uw\\ \rho Hu\end{bmatrix},\mathbf{F}=\begin{bmatrix}\rho v\\ \rho uv\\ \rho v^2+P\\ \rho vw\\ \rho Hv\end{bmatrix},\mathbf{G}=\begin{bmatrix}\rho w\\ \rho uw\\ \rho vw\\ \rho w^2+P\\ \rho Hw\end{bmatrix}$$

$$\mathbf{E}_v=\begin{bmatrix}0\\ \tau_{xx}\\ \tau_{xy}\\ \tau_{xz}\\ \varphi_x\end{bmatrix},\mathbf{F}_v=\begin{bmatrix}0\\ \tau_{yx}\\ \tau_{yy}\\ \tau_{yz}\\ \varphi_y\end{bmatrix},\mathbf{G}_v=\begin{bmatrix}0\\ \tau_{zx}\\ \tau_{zy}\\ \tau_{zz}\\ \varphi_z\end{bmatrix}$$

The elastic deformation of the lip is neglected for the variable geometry lip of the inlet. Only the rotation of the lip in the pitch direction along the rotating axis is considered. The motion characteristic parameters mainly include vibration amplitude and vibration frequency. The motion equation can be expressed as

$$\delta=\delta_0\sin(2\pi ft) \tag{2}$$

Where, δ_0 is the vibration amplitude of the lip pitch angle, f is the vibration frequency.

The deformation of the structure is determined by the vibration parameters of the lip. The dynamic grid technique is used to calculate the flow field, and the time step of the unsteady flow field is 1/50 of the vibration period of the lip.

4 RESULTS

4.1 *Computational model*

Assume that the corresponding deflection angle of lip water is 0°. Considering the movement of the inlet lip, a dynamic grid is used near the lip. During the CFD calculation, the grid is constantly adjusted with the structural deformation. In order to reduce the reconstruction time of the dynamic mesh, the dynamic mesh is only used near the deformation area, and the other areas are still fixed meshes. The final mesh model is shown in Figure 2.

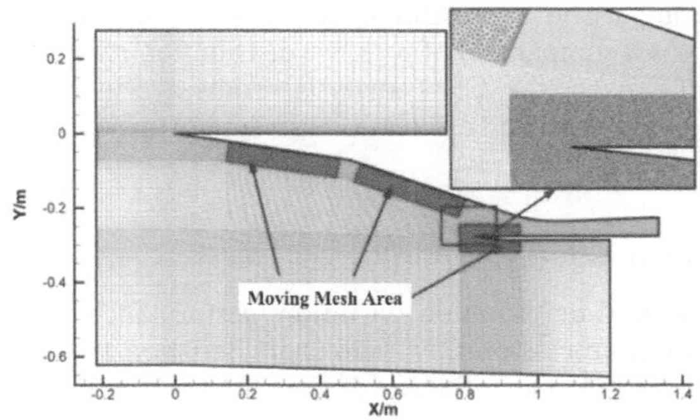

Figure 2. The mesh model of inlet.

In the calculation area, the dynamic grid area is unstructured grid, and other areas adopt structured grid. The second-order density-based implicit solver is used to solve CFD. The turbulence model is K-Omega SST model. The vector flux splitting scheme is AUSMDV.

4.2 *Verification of CFD analysis method*

The wind tunnel test data is obtained for typical airframe-propulsion integrated configuration. Figure 3 shows the distribution of pressure measuring points and the experimental model.

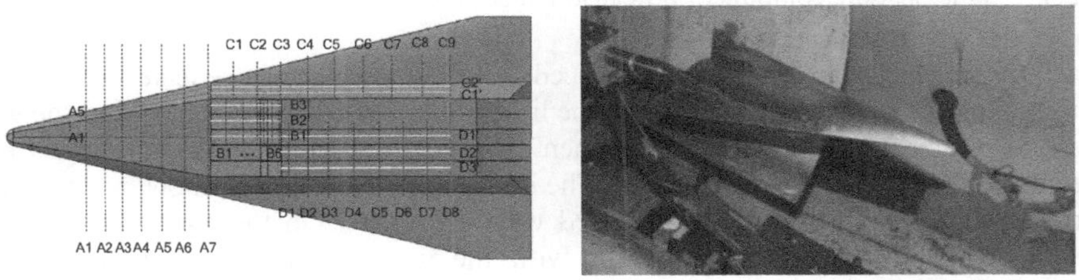

Figure 3. Wind tunnel test model.

At the state of Mach 6, and angle of attack is 0 degree. The pressure distribution along the fuselage is obtained based on wind tunnel experiment and CFD calculation. For the same incoming flow condition, the results comparison of calculation and test are as shown in Figure 4.

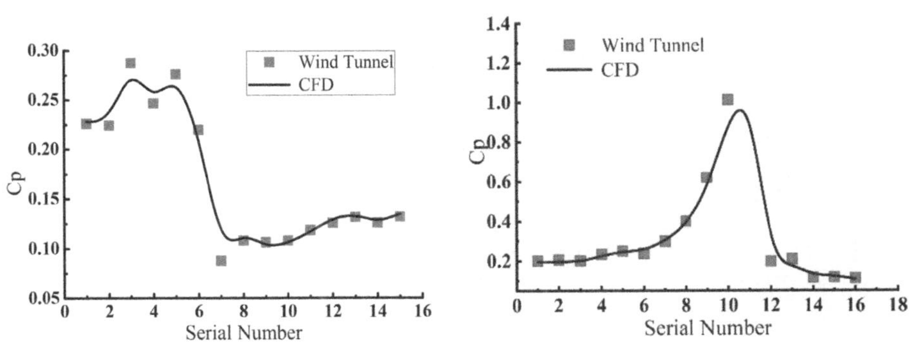

(a) (b)

Figure 4. Comparison between CFD and wind tunnel experiment.

Figure 4 (a) is a comparison of pressure distribution results on the side edge of the fuselage, and Figure 4 (b) is a comparison of pressure distribution results on the lower surface of the fuselage. By comparison, the CFD analysis results are in good agreement with the experimental results, and the related calculation can support performance impact analysis of the subsequent inlet.

4.3 *Results and analysis*

The influence of structural deformation on inflow performance is mainly reflected in the influence of deformation process on flow field characteristics. In this paper, only unidirectional coupling is considered. The influence of flow field on the structure is ignored.

In this section, the lip vibration of the inlet in the design state is studied. The calculated Mach number is 5, the height is 20km, and the calculated angle of attack is 0 degree.

4.3.1 *Influence of amplitude*

Assuming that the lip vibration frequency is 30Hz, the amplitudes are 1°, 3° and 5°, respectively. The static pressure contours of the flow field are shown in Figure 5 at the time of 0.2 s.

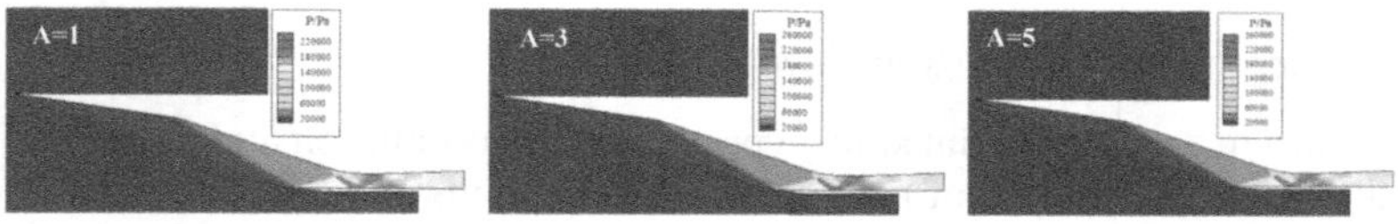

Figure 5. Static pressure contours of flow field under different lip vibration amplitudes.

A strong reflected shock wave is produced. The intensity of the reflected shock wave is determined by the relative angle between the lip and the airflow direction behind the wave of the second-stage external compression. When the lip deflects upward, the compression angle of the airflow behind the wave decreases. The intensity of the reflected shock wave decreases, and the static pressure behind the reflected wave increases. On the contrary, the pressure decreases, which is particularly obvious with the increase of the amplitude of the lip deflection angle.

As can be seen from Figure 6, the outlet performance appears sinusoidal oscillation with the lip vibration. The amplitude increases with the increase of lip vibration amplitude. When the lip amplitude is 1 °, 3 ° and 5 °, the static pressure amplitude is 2.06%, 3.09% and 10.8%

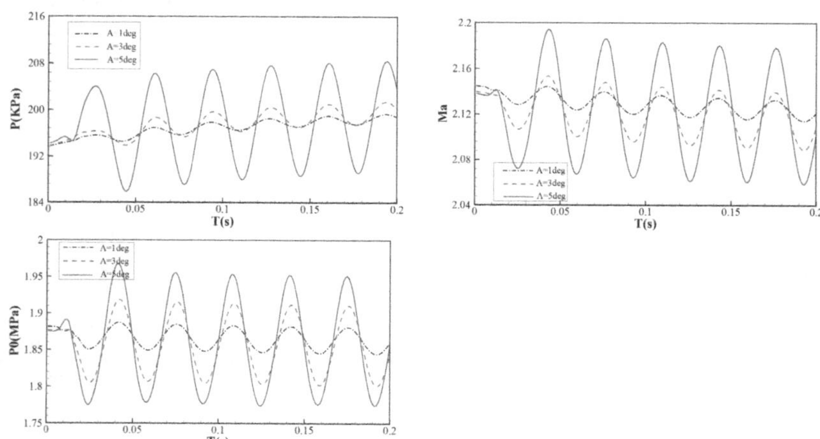

Figure 6. Oscillation curve of outlet performance.

of the initial values, Mach number amplitude is 0.9%, 1.45% and 5.6% of the initial values, total pressure amplitude is 5.6%, 2.16%, 5.85% and 9% of the initial values. When the lip amplitude increases, the static variation amplitude of the inlet increases sharply. However, the total pressure and other parameters are almost consistent with the lip amplitude. The lip vibration mainly affects the oscillation of inlet's outlet parameters, but has limited influence on its average performance. In 0.2 s vibration time, the average static pressure increases by about 3.1%, the Mach number decreases by about 2.3%, and the total pressure loss is about 0.53%. The change of outlet average performance is almost the same under different lip amplitude.

4.3.2 *Influenc of frequency*

Assuming the lip amplitude of the inlet is 3 degrees, the static pressure contours of the inlet is shown in Figure 7 at the time of 0.2 s. The vibration frequencies are 10Hz, 30Hz and 50Hz, respectively.

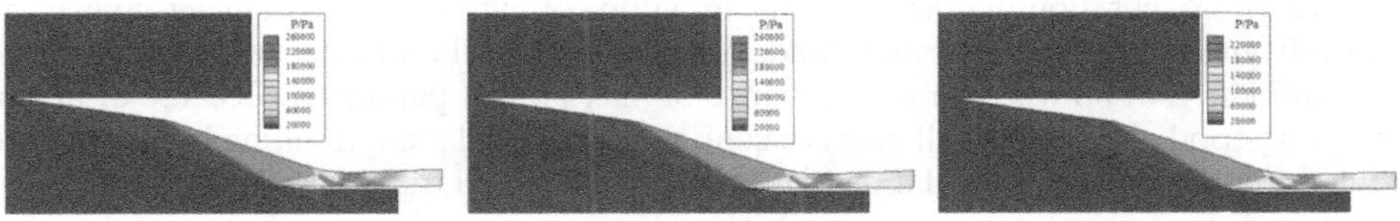

Figure 7. Static pressure contours of flow field under different lip vibration frequencies.

When the lip amplitude is the same, the vibration frequency of the inlet mainly affects the internal flow oscillation. However, the influence on the flow field structure is limited. Different vibration frequencies will not destroy the structure of the inlet flow field, so the features of the wave reflection are the same.

The curves of inlet's outlet flow parameters changing with time are shown in Figure 8. Due to the influence of vibration frequency, the oscillation frequency of inlet's outlet parameters is the same as that of lip. The oscillation amplitude of the parameters is quite different

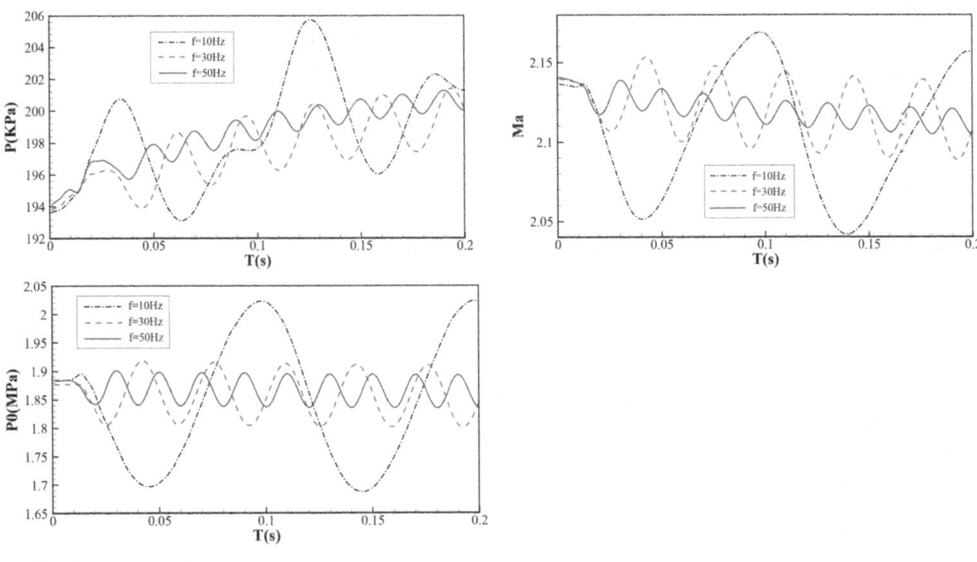

Figure 8. Oscillation curve of inlet outlet performance.

under the same lip vibration amplitude. The amplitude of the parameters increases greatly with the decrease of lip vibration frequency. At the same amplitude (3 °), the amplitude of static pressure at different vibration frequencies (10Hz, 30Hz and 50Hz) accounts for 1.03%, 2.06% and 5.15% of the initial value, respectively. The amplitude of Mach number at different vibration frequencies accounts for 0.7%, 2.3% and 5.6% of the initial value, respectively. The amplitude of total pressure at different vibration frequencies accounts for 2.7%, 5.4% and 17.3% of the initial value, respectively. Lower vibration frequency leads to large amplitude vibration of inlet's outlet parameters, which is mainly due to the large influence of lip angle on inlet internal flow performance. When the vibration frequency is low, the influence lasts for a long time, resulting in large amplitude change of inlet's outlet parameters. When frequency vibration is high, the continuous accumulation time of internal flow performance change caused by lip deformation is short in a vibration cycle. The oscillation frequency of final outlet parameters is high but the amplitude decreases.

Based on the unsteady flow field, the oscillation amplitude of the inlet's outlet parameters is almost consistent with that of the lip vibration under the same vibration frequency. However, the effect of lip vibration frequency on the amplitude of inlet's outlet parameter oscillation is obviously greater than vibration amplitude of lip. When lip vibration occurs at low frequency, the deformation of lip will lead to the change of inlet's outlet parameter. Because of the long vibration period, the change will produce continuous accumulation, resulting in the amplitude of inlet's outlet parameter oscillation higher than that of high frequency oscillation.

5 CONCLUSION

The influence of lip geometry adjustment on inlet performance is analyzed by CFD. The influence of various motion factors on inlet's outlet performance is emphatically compared. The related research shows that the inlet's outlet performance is often not a simple harmonic oscillation when the inlet structure vibrates harmonically. The reason is the continuous cumulative effect of the flow field. With the accumulation of vibration time, the average performance of the inlet's outlet gradually changes. Relevant research conclusions include:

(1) Under the same frequency, the larger the amplitude of the lip, the greater the influence on the inlet's outlet performance. The influence relationship is not a simple linear relationship.
(2) Under the same amplitude, the vibration frequency of the lip not only affects the oscillation frequency of the outlet performance, but also affects the oscillation amplitude of the outlet performance. It has a greater influence on the oscillation amplitude.

ACKNOWLEDGEMENT

This work was financially supported by National Numerical Wind Tunnel.

REFERENCES

Adam J. Culler and Jack J. McNamara. (2010). Studies on fluid–thermal–structural coupling for aerothermoelasticity in hypersonic flow [J]. *AIAA Journal*, , 48(8):1721–1738.

Andrew D. Clark, Maj Dean Mirmirani *et al.* (2006). An aero-propulsion integrated elastic model of a generic airbreathing hypersonic vehicle [C]. *AIAA Guidance, Navigation and Control Conference and Exhibit, Keyston*. Colorado.

Baris Fidan, Matthew Kuipers, and Petros A. Ioannou. (2006). Longitudinal motion control of air-breathing hypersonic vehicles based on time-varying models[C]. *American Institute of Aeronautics and Astronautics 14th AIAA/AHI Space Planes and Hypersonic Systems and Technologies Conference*. Canberra.

Bing Chen, Hongyang Chen, Tao Mo *et al.* The analysis on the flow characteristics of supersonic inlet considering structural vibration[C], *Journal of Physics: Conference Series, FDCS-2022*: 1–10.
CHEN Bing, GU Liangxian, GONG Chunlin. (2013). Study on variable geometry inlet of acceleration hypersonic vehicle [J]. *Journal of Solid Rocket Technology*, 36(4): 431–436.
David K. Schmidt. (1992). Dynamics and control of hypersonic aeropropulsivel/aeroelastic vehicles [C]. *American Institute of Aeronautics and Astronautics Guidance, Navigation and Control Conference,* Hilton Head Island.
Edelfelt I.H. (1955). Variable Throat Supersonic Diffuser, *America, 3011307[P].*
Frank R. Chavez and David K. (1994). Schmidt. Analytical aeropropulsive/aeroelastic hypersonic vehicle model with dynamic analysis[J]. *Journal of Guidance Control and Dynamics*, 17(6): 1308–1319.
Frank R. Chavez and Dr. David K. Schmidt. (1992). An integrated analytical aeropropulsivel aeroelastic model for the dynamic analysis of hypersonic vehicles[C]. *American Institute of Aeronautics and Astronautics Guidance, Navigation and Control Conference,* Hilton Head Island.
Frank R. Chavez and Dr. David K. Schmidt. Dynamics of hypersonic flight vehicles exhibiting significant aeroelastic and aeropropulsive interactions[C]. *American Institute of Aeronautics and Astronautics Guidance, Navigation and Control Conference*, 1993. Monterey.
Hui MA, Jian-Ping YUAN and Qun FANG. (2006). An analytical model of dynamics for generic airbreathing hypersonic vehicles[C]. *American Institute of Aeronautics and Astronautics 14th AIAA/AHI Space Planes and Hypersonic Systems and Technologies Conference.* Canberra.
Jack J. McNamara *et al.* (2008). Aeroelastic and Aerothermoelastic Behavior in Hypersonic Flow [J]. *AIAA Journal,* 2008, 46(10):2591–2612.
James Moss, Dr Eric Perrell. (2004). Design and preliminary analysis of variable geometry SCRAM jet/ Rocket[C], *40th AIAA/ASME/SAE/ASEE Joint Propulsion Conference and Exhibit.* Florida.
JIN Zhiguang, ZHANG Kunyuan. (2010). A Variable Geometry Scramjet Inlet with a Translating Cowl Operating in a Large Mach Number Range[J]. *Journal of Astronautics.* 31(5):1503–1519.
John D. Anderson. (2009). Computational fluid dynamics [M]. *China Machine Press.*
Kline H, Palacios F, Alonso J J. (2014). Sensitivity of the performance of a 3-dimensional hypersonic inlet to shape deformations[C]. *19th AIAA Space Planes and Hypersonic Systems and Technologies Conference,* Atlanta.
YE Kun, YE Zheng-yin, QU Zhan. (2017). Effects of panel vibration on performance of combustor for scramjet [J]. *Journal of Propulsion Technology*, 38(2):386–398.

Fluid Mechanics and Industrial Applications – Fang (Ed.)
© 2026 The Author, ISBN: 978-1-041-15371-9

Research on thermodynamic calculation correction method for phase adjustment and pressure water process of pump turbine

Hengle Yang and Hao Zhang
Henan Tianchi Pumped Storage Co., Ltd, Henan, China

Xia Liu
Zhongnan Engineering Corporation Limited, Changsha, China

Qingnan Li
Henan Tianchi Pumped Storage Co., Ltd, Henan, China

ABSTRACT: During the process of phase regulation and pressurized water supply, the fluid temperature inside the gas supply system of the pumped storage power station's water pump turbine unit undergoes significant changes. The drastic changes in temperature will have a certain impact on the strength of equipment and pipelines, posing safety hazards during operation. During the design process, the volume of the gas storage tank is estimated and selected based on relevant standards and specifications. The fluid temperature inside the gas storage tank after pressurized water is calculated through thermodynamic theory. During the calculation process, the correlation coefficients were taken based on empirical values, and certain assumptions were made about the water pressure process and gas state, resulting in low accuracy of the final temperature values. This article is based on on-site measured data and converts the internal fluid temperature distribution of the system through heat transfer formulas. The measured data is used to correct the thermodynamic theoretical calculation of the pressurized water supply process, and the gas volume and correlation coefficient values are derived. The gas volume obtained based on measured data and the further selection of gas storage tank volume reasonably consider the influence of temperature, which can reflect the actual operation of the unit. The proposed correlation coefficient value is also an exploratory study of theoretical design values and can provide reference for the design of other similar power plants.

1 INTRODUCTION

During the phase regulation and pressurized water process, works due to the expansion of high-pressure compressed gas will cause a sudden drop in fluid temperature in the gas supply system of the pump turbine unit. The drastic temperature changes will have a certain impact on the service life and safe operation of equipment and pipelines in the system. In the design process, the temperature change inside the gas storage tank is usually calculated based on certain assumptions and thermodynamic formulas (Lei Yang, 2016). The parameters involved in the analysis process were in accordance with relevant regulatory requirements, without considering the influence of temperature. Therefore, the thermodynamic analysis process of pressurized water supply during the design process is relatively simple and rough. During actual operation, the heat exchange between the gas supply system and the outside world, the material characteristics of equipment and pipelines, and the physical and state parameters of the internal fluid all have a significant impact on the temperature field over

DOI: 10.1201/9781003679561-5

time. There may be significant differences between the actual temperature of the gas in the gas storage tank during the pressurized water process and the theoretical calculation results. The calculation of gas storage tank volume under actual temperature conditions will be more reasonable for the design and selection of power plant gas supply systems.

This article uses on-site measurement methods to collect temperature data throughout the entire system during the pressurized water supply process, compares it with theoretical calculation results, re estimates the volume of compressed gas, and proposes recommended values for relevant parameters in the design and selection process.

2 THERMODYNAMIC THEORETICAL CALCULATIONS

This article takes the phase-shifting pressurized water supply system of a pumped storage power station as an example. According to the selection requirements for the storage tank volume of the pressurized water supply system in the "Design rule of hydraulic mechanical auxiliary equipment system of hydraulic power plant" (NB/T35035), the storage tank volume can be calculated using formula (1). The results are detailed in Table 1. This article takes the actual operating tailwater pressure, working pressure of the air storage tank, and inflation volume, with a variable value of n taken as 1.2. The inflation volume includes the amount of air leakage. The calculated volume of the gas storage tank V_t is 13.59m^3. The power station currently selects two 8m^3 gas storage tanks, which can meet the design requirements.

$$V_t = \frac{(P_d + 1)^{\frac{1}{n}}}{(P_r + 1)^{\frac{1}{n}} - (P_r' + 1)^{\frac{1}{n}}} V_d(N + \alpha) \tag{1}$$

In the formula, V_t is the gas storage tank capacity, m^3; P_d is the air pressure inside the tailpipe, kgf/cm^2; P_r is the lower limit of normal working pressure of gas storage tank, kgf/cm^2; P_r' is after completing one water pressure operation, the pressure in the gas storage tank is the lowest allowable pressure inside the tank, kgf/cm^2, 0.3MPa higher than the pressure inside the tailpipe; V_d is the inflation volume inside the impeller chamber and draft tube, m^3; n is the polytropic index, taking $n \geq 1.2$; N is the the number of pump turbines that need to lower the water surface; α is the number of backup water pressure operations (α is 0.5-1). For the unit method, take (α+1) times, with a short interval of 1 for α, and a long interval of 0.5 for α.

Table 1. Calculation of storage tank volume.

P_d kgf/cm^2	P_r kgf/cm^2	P_r' kgf/cm^2	V_d m^3	N	n	α	V_t m^3
11.53	79.04	14.59	15.81	1	1.2	1	13.59

When conducting thermodynamic calculations, assume the following:

(1) Compressed air is considered an ideal gas during the water pressure process.
(2) During the water pressure process, there is no heat exchange between the gas storage tank and the external environment.
(3) During the water pressure process, there is no air leakage from the air storage tank, and the system quality is conserved.

The water pressure process is relatively fast. Therefore, based on the above assumption, the change of compressed air inside the gas storage tank during the water pressure process is considered as an isentropic adiabatic process. Then follow the following formula, where the air insulation coefficient r is taken as 1.4.

$$PV = mRgT \tag{2}$$

$$\frac{T_n}{T_0} = \left(\frac{P_n}{P_0}\right)^{\frac{r-1}{r}} \tag{3}$$

In the formula, P is the compressed air pressure, Pa; footmark 0 represents the initial state, and n represents the end state; V is the air volume, m^3; m is the air mass, kg; Rg is the air gas constant, 287J/(kg·K); T is the air temperature, K; r is the air insulation coefficient.

According to formula (2), taking the ambient temperature of the tailwater layer as 14.40 °C and the ambient temperature of the gas storage tank layer as 18.33 °C, the required compressed air mass m_d for one water pressurization can be calculated as 216.46kg. The compressed air mass m_0 inside the gas storage tank before pressurization is 1258.40kg, as shown in Tables 2 and 3. The remaining air mass m_1 in the air storage tank after one water pressurization can be calculated by m_1=m_0-m_d, which is 1041.94kg. Similarly, the remaining mass m_2=m_1-m_d in the gas storage tank after two consecutive water pressurizations can also be obtained, which is 825.49kg.

Table 2. Gas mass calculation during a water pressure process.

P_d MPa	V_d m3	Rg J/(kg·K)	T_d K	m_d kg
0.968	15.81	287	287.55	216.46

Table 3. Calculation of gas mass inside gas storage tanks during a water pressure process.

P_0 MPa	V_0 m^3	Rg J/(kg·K)	T_0 K	m_0 kg
7.746	13.59	287	291.48	1258.40

By jointly solving equations (2)and (3), equations (4) and (5) can be obtained. m_1 has been obtained, and T_1 is −2.87°C and P_1 is 5.947MPa. Similarly, it can be concluded that T_2 is −26.91 °C and P_2 is 4.293 MPa.

$$T_1 = \left(\frac{m_1 Rg}{P_0 V}\right)^{r-1} T_0^r \tag{4}$$

$$P_1 = \left(\frac{T_1}{T_0}\right)^{\frac{r}{r-1}} P_0 \tag{5}$$

3 ON SITE MEASUREMENT

3.1 *Layout of measuring points*

Paste temperature strain gauges on the surface of gas supply system equipment and pipelines, collect real-time signal data, and obtain the temperature changes of the gas supply system during the water pressure process. Six measuring points Pt1 to Pt6 are arranged along the gas supply direction, namely the junction between the gas storage tank and the gas supply pipeline, in front of the gas supply main valve, behind the gas supply main valve, in front of the throttle orifice plate, behind the throttle orifice plate, and at the junction between the gas supply pipeline and the tailwater cone pipe, as shown in Figure 1.

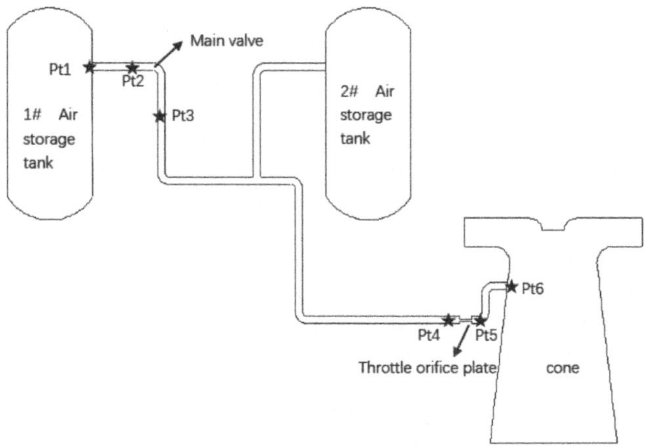

Figure 1. Schematic diagram of measuring points.

3.2 *Test data*

The on-site measurement of water pressure time was 28s. During the actual measurement of water pressure, the pressure of the gas storage tank decreased from 7.746 MPa to 5.553 MPa. The temperature measurement data is shown in Table 4. The results showed that the temperature change at Pt4 measuring point after the throttle orifice plate was the most severe, with a change of up to 32.17 °C after two consecutive water pressures, and the final temperature had dropped to −20.47 °C. Next is the Pt3 measuring point behind the main gas supply valve, with a temperature drop of 19.79 °C after two consecutive water pressure tests. Next, the temperature at Pt2 in front of the gas supply main valve, Pt3 in front of the throttle orifice plate, Pt1 at the junction of the gas storage tank and the gas supply pipeline, and Pt5 at the junction of the gas supply pipeline and the tailwater cone pipe remains almost unchanged.

Table 4. Pressure water temperature test results (unit: °C).

Point	Pt1	Pt2	Pt3	Pt4	Pt5	Pt6
Start	18.33	17.60	15.30	13.58	11.70	13.25
End of one water pressure test	15.89	8.08	3.36	8.44	−18.14	13.17
One temperature drop	2.44	9.52	11.94	5.14	29.84	0.08
End of two water pressure cycles	9.75	0.05	−4.49	3.49	−20.47	12.95
Two temperature drops	8.58	17.55	19.79	10.09	32.17	0.30

3.3 *Test data correction*

According to heat transfer theory, the heat flux between the external environment and the outer wall, between the outer wall and the inner wall, and between the inner wall and the internal fluid is the same in a stable state. The temperature of the external wall pipeline collected during testing can be converted into the internal fluid temperature through equations 6 to 8.

$$t_{wo} - t_{fo} = \frac{\phi}{h_o \pi d_o l} \tag{6}$$

$$t_{wi} - t_{wo} = \frac{\phi}{2\pi\lambda l} \ln \frac{d_0}{d_i} \tag{7}$$

$$t_{fi} - t_{wi} = \frac{\phi}{h_i \pi d_i l} \tag{8}$$

In the formula, t_{fi} is the internal fluid temperature, °C; t_{wi} is the inner wall temperature, °C; t_{wo} is the outer wall temperature, °C; Φ is the heat flux, W; hi is the heat transfer coefficient between fluid and inner wall, W/(m^2·K); λ is the thermal conductivity coefficient, W/(m·K); d_o is the tube outer diameter, m; d_i is the tube inner diameter, m; l is the feature length, m.

The calculation of the heat transfer coefficient h_o between the external environment and the external wall of the pipe, as well as the heat transfer coefficient h_i between the fluid inside the pipe and the internal wall, is crucial. The heat transfer coefficient reflects the intensity of heat transfer, and its magnitude not only depends on the fluid participating in heat transfer, but also on processes such as flow rate and phase change. The calculation is shown in Equation (9). Among them, the Nusselt number Nu is a dimensionless number that reflects the flow state and thermal diffusion ability of the fluid.

$$h = Nu\frac{\lambda}{l} \tag{9}$$

In the formula, h is the heat transfer coefficient, W/(m^2·K); Nu is the Nusselt number; λ is the thermal conductivity coefficient, W/(m·K); l is the feature length, m.

For the heat transfer of fluid outside the pipe, the Nusselt number Nu can be obtained by equations (10) to (12). Among them, the coefficients C and n are both related to the surface shape, position, and flow state, and are selected based on the calculated Gr number.

$$Nu_m = C(Gr\mathrm{Pr})_m^n \tag{10}$$

$$Gr = \frac{g\alpha_V \Delta t l^3}{\nu^2} \tag{11}$$

$$\mathrm{Pr} = \frac{\nu}{a} = \frac{c_p\mu}{\lambda} \tag{12}$$

In the formula, Nu is the Nusselt number; Gr is the Grayshev number, buoyancy/viscous force; Pr is the Prandtl number, a comparison of flow diffusion and thermal diffusion abilities in fluids; C and n is the coefficient; g is the gravitational acceleration, m/s^2; α_V is the coefficient of body expansion, 1/T, K^{-1}; Δt is the temperature difference, °C; l is the feature length, m; ν is the kinematic viscosity, $\nu = \mu/\rho$, m^2/s; a is the thermal diffusivity, m^2/s; c_p is the specific pressure heat capacity, J/(kg·K); μ is the dynamic viscosity, Pa·s; λ is the thermal conductivity coefficient, W/(m·K).

For convective heat transfer inside a tube, the Nusselt number Nu can be calculated using the Gnielinski formula (Gnielinski V, 1976) (Gnielinski V, 2015), as detailed in equations (13) to (14). This formula is the most accurate correlation formula to date.

$$Nu_f = \frac{(f/8)(\mathrm{Re} - 1000)\mathrm{Pr}_f}{1 + 12.7\sqrt{f/8}\left(\mathrm{Pr}_f^{2/3} - 1\right)}\left[1 + \left(\frac{d}{l}\right)^{2/3}\right] \tag{13}$$

$$f = (1.8 \lg \mathrm{Re} - 1.5)^{-2} \tag{14}$$

In the formula, Nu is the Nusselt number, footmark f represents qualitative temperature; f is the darcy drag coefficient of turbulent flow in pipes; Re is the Reynolds number; Pr is the Prandtl number; d is the tube inner diameter; l is the feature length, m; ν is the kinematic viscosity, m^2/s; a is the thermal diffusivity, m^2/s; c_p is the specific pressure heat capacity, J/(kg·K); μ is the dynamic viscosity, Pa·s; λ is the thermal conductivity coefficient, W/(m·K).

The outer diameter and inner diameter of the pipeline in this gas supply system are 108mm and 92mm, and all physical parameters are queried based on qualitative temperature and corresponding metal materials (Wenquan Tao, 2019). By calculating, the heat transfer coefficient and heat flux of each measuring point at the end of a water pressure test can be obtained, and the internal fluid temperature can be converted, as shown in Table 5. The inner wall temperature is almost the same as the outer wall temperature, and the internal fluid temperature is slightly lower than the outer wall temperature. For the two measuring points of the gas storage tank and the tailwater cone, the heat transfer coefficient between the internal fluid and the inner wall is relatively small compared to other measuring points, and the maximum difference in temperature between the internal fluid and the outer wall is about 1 °C. At the measuring points on the pipeline, the heat transfer coefficient between the internal fluid and the inner wall is above 1500 W/(m^2·K), and there is almost no difference in the temperature of the internal fluid and the temperature of the inner and outer walls.

The conversion results of internal fluid temperature for two consecutive water pressure processes are shown in Table 6. After two consecutive water pressures, the fluid temperature inside each measuring point further decreases. The temperature at measuring point Pt1 of the gas storage tank decreased to 4.07 °C; There is a temperature difference between Pt2 and Pt3 measuring points before and after the main gas supply valve, with the lowest temperature being -0.06 °C and -4.68 °C, respectively. The influence of the valve results in a temperature difference of 4.62 °C; The temperature difference between points Pt4 and Pt5 before and after the throttle orifice plate is greater, with the lowest temperature being 3.43 °C and -20.72 °C, respectively, with a difference of 24.15 °C; The temperature change at the Pt6 point of the tailwater cone is very small, which is basically consistent with before water pressure. Analysis suggests that the setting of the gas supply main valve and throttle orifice plate has a significant impact on the fluid velocity inside the pipeline, especially the throttle orifice plate. The fluid was originally flowing in a pipeline with an inner diameter of 92mm, but suddenly passed through an orifice plate which diameter is 22mm, causing a sudden increase in velocity and a further decrease in temperature. This position is the most important point to pay attention to in the entire gas supply system.

Table 5. Calculation results of internal fluid temperature at the end of a water pressure test.

Point	Unit	Pt1	Pt2	Pt3	Pt4	Pt5	Pt6
t_{fo}	°C	18.33	18.33	18.33	14.40	14.40	14.40
t_{wo}	°C	15.89	8.08	3.36	8.44	−18.14	13.17
t_m	°C	17.11	13.20	10.85	11.42	−1.87	13.79
Pr		0.70	0.70	0.70	0.70	0.70	0.70
Gr		2.60e+10	2.21e+06	3.22e+08	1.29e+06	7.39e+06	5.29e+09
C		0.11	0.48	0.59	0.48	0.48	0.03
n		0.33	0.25	0.25	0.25	0.25	0.39
Nu_m		289.64	16.95	72.41	14.82	22.93	168.88
h_o	W/(m^2·K)	1.81	3.94	3.64	3.44	5.33	1.36
Φ	W	100.64	32.88	9.23	13.94	117.66	34.92
t_{wi}	°C	15.89	8.05	3.33	8.43	−18.24	13.17
h_i	W/(m^2·K)	4.31	1556.29	339.68	1575.35	1563.39	5.60
t_{fi}	°C	14.83	8.02	3.14	8.41	−18.37	12.86

Table 6. Calculation results of internal fluid temperature at the end of two consecutive water pressure tests.

Point	Unit	Pt1	Pt2	Pt3	Pt4	Pt5	Pt6
t_{fo}	°C	18.33	18.33	18.33	14.40	14.40	14.40
t_{wo}	°C	9.75	0.05	−4.49	3.49	−20.47	12.95
t_m	°C	14.04	9.19	6.92	8.94	−3.04	13.68
Pr		0.70	0.70	0.70	0.70	0.70	0.70
Gr		9.22e+10	3.99e+06	4.98e+08	2.38e+06	7.95e+06	7.45e+09
C		0.11	0.48	0.59	0.48	0.48	0.03
n		0.33	0.25	0.25	0.25	0.25	0.39
Num		441.83	19.65	80.75	17.28	23.35	180.37
h_o	W/(m²·K)	2.76	4.57	4.05	4.02	5.43	1.45
Φ	W	539.18	67.98	15.69	29.73	128.42	44.14
t_{wi}	°C	9.74	0.00	−4.54	3.46	−20.58	12.95
h_i	W/(m²·K)	4.29	1549.47	337.97	1572.12	1561.64	5.60
t_{fi}	°C	4.07	−0.06	−4.86	3.43	-20.72	12.55

4 DISCUSSION ON THERMODYNAMIC CALCULATION AND CORRECTION

The comparison between on-site measurement and theoretical calculation results is shown in Table 7. Comparing the pressure values, the measured and calculated pressure values at the end of one and two water pressure tests are in good agreement, with a maximum deviation of 6.52%. By comparing the temperature values and conducting on-site measurements and conversions, the fluid temperature inside the gas storage tank reached 14.83 °C at the end of one water pressure test and 4.07 °C at the end of two consecutive water pressure tests; The corresponding temperature calculated theoretically is −2.87 °C, which is significantly different from −26.91 °C. Analysis suggests that the theoretical calculation assumes that compressed air is an ideal gas, and there is no heat exchange between the gas storage tank and the external environment. The change of compressed air inside the gas storage tank during the water pressure process is considered an isentropic adiabatic process. However, the compressed gas pressure is relatively high and cannot be approximated as an ideal gas. Moreover, heat exchange always occurs, and the working conditions of the gas storage tank are non-adiabatic, which leads to a deviation between the measured data and the theoretical data. Under actual conditions, according to the maximum temperature drop, the temperature change of the gas storage tank decreased by 14.26 °C, while the theoretically calculated temperature change reached 45.24 °C.

Table 7. Comparison of results between actual measurement and theoretical calculation of gas storage tanks.

Parameter	Initial pressure P_0	End pressure of one water pressure test P_1	End pressure of two water pressures P_2	Initial temperature T_0	End temperature of one water pressure test T_1	End temperature of two water pressures T_2
Unit	MPa	MPa	MPa	°C	°C	°C
Actual measured value	7.746	5.583	4.384	18.33	14.83	4.07
Calculated value	7.746	5.947	4.293	18.33	−2.87	−26.91

This article intends to use measured data to correct the volume of compressed gas and the polytropic index n. According to formulas (2), formulas (15) and (16) can be derived. According to the temperature calculation at the end of a water pressure test, the gas volume V = 8.64m^3 and the initial gas mass m_0 = 800.26kg can be obtained. Due to the constant gas mass md required to complete water pressure, m_1 = 583.81kg. In theoretical calculations, the gas mass required for two consecutive water pressurizations is directly simplified to $2m_d$. However, in actual testing, the second water pressurization process is not allowed after the first water pressurization is completed. Instead, the exhaust valve is opened to allow the tail water to partially reflux, and the second water pressurization is carried out when the pressure in the gas storage tank returns to a fixed state. Therefore, based on the measured data of two consecutive water pressure tests, according to formulas (17), it can be calculated that m_2 = 476.22kg, which means that the actual gas mass required for the second water pressure test is Δm = 107.58kg, which is significantly different from the theoretical calculation of the second water pressure test using the value of md. This is also one of the reasons for the difference between the theoretical calculation and the measured temperature results.

$$V = \frac{P_d V_d}{T_d} \times \frac{T_0 T_1}{P_0 T_1 - P_1 T_0} \tag{15}$$

$$m_0 = \frac{P_0 T_1}{P_0 T_1 - P_1 T_0} m_d \tag{16}$$

$$m_2 = \frac{P_2}{P_1} \times \frac{T_1}{T_2} \times m_1 \tag{17}$$

Based on measured data, the air insulation coefficient r=0.96 can be calculated using formulas (3). According to V=8.64m^3 and formula (1), the polytropic index can be obtained as approximately n=0.97. According to thermodynamic knowledge, the polytropic index can usually vary within any value range. Taking the derivative of the polytropic process state equation (Yingzi Jin, 2005) yields formulas (18). During the venting process of high-pressure vessels, if the pressure and density always change, then n is also constantly changing. It can be said that in a short period of time when the pressure and density remain constant, the variable index n remains unchanged. Therefore, to calculate the value of n in detail, the entire process of pressurized water supply needs to be divided into several stages for analysis. This article simplifies the estimation of the variable index of the entire gas supply process based on the temperature measurement data at the initial and end times of the pressurized water supply process, and only discusses the revision of the previous theoretical calculation method for estimating the volume of the pressurized water supply gas storage tank in water pump turbines. To further accurately propose recommended values for the polytropic index and adiabatic coefficient, it is necessary to accumulate a large amount of measured data and calculate according to statistical laws. The suggested values presented in this article are only based on the data from this experiment. The recommended value of the variable index n in standard NB/T35035 is relatively large, which is beneficial for ensuring the volume of the gas storage tank. However, according to the measured data in this study, the temperature drop during the actual operation of the gas storage tank is not as large as the theoretical calculation. Therefore, when selecting low-temperature materials for gas storage tanks, reference can be made.

$$\frac{dp}{dv} = -\mathrm{n}\frac{p}{v} \tag{18}$$

In the formula, v is the specific volume, m^3/kg; p is the pressure, Pa; n is the polytropic index.

5 CONCLUSION

This article uses on-site measurement methods to obtain the degree and distribution of fluid temperature changes inside the gas supply system of the pumped storage power station pump turbine unit during the phase regulation and water pressure process.

Along the gas supply direction, in the entire gas supply system, the setting of the throttle orifice device on the pipeline causes the most drastic temperature changes at the subsequent measuring points, which can be used as a key monitoring location for actual operation. The temperature of the gas storage tank also significantly decreased after being pressurized with water, and the temperature decreased by about 14 °C for two consecutive times of pressurized water.

The actual water pressure process during the operation of the unit is different from the theoretical calculation under certain assumptions. The actual measured temperature drop on the gas supply system equipment and pipelines is relatively small compared to the theoretical calculation. After two consecutive water pressure tests on the air storage tank, the calculated temperature is −26.91 °C, while the actual measurement is 4.07 °C.

Based on the actual measurement data, the volume and variability index of the gas supply were revised and estimated, which were relatively small compared to the values calculated according to the standard in the past. The theoretical calculation value of the gas supply volume is $13.59m^3$, and the corrected calculation value is $8.64m^3$; The theoretical calculation value of the polytropic index is 1.2, and the corrected calculation value is 0.97.

ACKNOWLEDGMENTS

This work was financially supported by the State Grid Xinyuan Company LTD Science and Technology Project fund (SGXYKJ-2022-074).

REFERENCES

Gnielinski V, (1976). New equations for heat mass transfer in turbulent pipe and channel flows, *Int Chem Eng*, 16:359–368.

Gnielinski V, (2015). On heat transfer in tubes, *Int Journal Heat Mass Transfer*, 63: 134–140.

Lei Yang, Jianxing Zheng, (2016). Thermodynamic analysis and discussion on the phase regulation and pressurized water process of the Heimifeng pumped storage power station, *Mechanical & Electrical Technique of Hydropower Station*, 2016(39), 2: 61–65, 80.

Wenquan Tao, (2019). *Heat Transfer*, Higher Education Press, Beijing, pp. 529–559.

Yingzi Jin, Zuchao Zhu, (2005). Simplification and determination of polytropic exponent of thermodynamic process in the filling and exhausting process in a pneumaitic system, *Chinese Journal of Mechanical Engineering*, 2005(41), 6: 76–79.

Fluid Mechanics and Industrial Applications – Fang (Ed.)
© 2026 The Author, ISBN: 978-1-041-15371-9

Numerical study of the impact of temperature on an axial fan performance

Haimeng Huang, Changde Li and Guangliang Tang
Guangdong Zhaoli Motor Group Co., Ltd., Zhongshan, Guangdong, China

ABSTRACT: Industrial fans are used in various fields such as electric fans, computer coolers, air fryers, air conditioners, and more. The temperature range of the computer cooler is between 20°C and 90°C, and the temperature range of the air fryer is between 20°C and 200°C, and the temperature range of an aeroengine is between 20°C and 2000°C. Ambient temperature is also taken into account and the extreme minimum temperature is -53°C at Mohe City during winter. However, research on the impact of temperature on fan performance has not yet been conducted. Here, we study the impact of temperature on an axial fan performance by ANSYS Fluent. The temperature ranges of -40°C to 200°C and -50°C to 1200°C were studied separately. The numerical results show that temperature plays a crucial role in the axial fan and the flow rate of outlet and the power of impeller decrease as the temperature increases in the whole. For examples, the corresponding flow rate of outlet decreases to the initial 81% as the computer cooling fan heats up from 20°C to 90°C, and the initial 61% as the air fryer fan heats up from 20°C to 200°C, and the corresponding flow rate of outlet decreases to the initial 19% as the aircraft engine heats up from 20°C to 1200°C. Those results support the conclusion that temperature plays a key role in the axial fans, and the design and numerical simulation of axial fan must take into account the impact of temperature.

1 INTRODUCTION

Axial fans are most widely used in civilian, military, and industrial fields (such as computer cooler and air fryers) due to their good economic efficiency, stable structure, high efficiency, and large flow rate. The operating temperature range of axial fans primarily focuses on 0°C ~200°C in household appliances. For examples, the temperature range of the computer cooler is between 20°C and 90°C, and the temperature range of the air fryer is between 20°C and 200°C, and the operating temperature range of the electric fan heater is between 20°C and 150°C. The operating temperature of an aeroengine can reach up to 2000°C. In further, ambient temperature is also taken into account and the extreme minimum temperature is −53°C at Mohe City during winter.

In recent years, considerable work has been conducted on the design of axial fans, primarily driven by the pursuit of high aerodynamic performance. An optimization program for axial fan design was developed, which increased fan efficiency by 1.8% and achieved a 31.4% improvement (Antoine Moreau, 20211). Sarraf and his team studied the impact of blade thickness on the performance of axial fans with controlled vortex design, and they concluded that thick blades have lower aerodynamic efficiency than thin blades (C. Sarraf, 2011). Spasic *et al.* designed and tested the performance of reversible fans with two different blade configurations, and found that compared to fans with straight-line blade profiles, reversible axial fans with slightly hyperbolic blade profiles exhibited greater improvements in pressure and flow rate (Zivan Spasic, 2018).

DOI: 10.1201/9781003679561-6

However, research on the impact of temperature on the axial fan performance has not yet been conducted. Here, we study the impact of temperature on an axial fan performance by numerical simulation. The temperature ranges of -40°C to 200°C and -50°C to 1200°C were studied separately. The numerical results show that temperature plays a crucial role in the axial fans.

2 NUMERICAL METHOD

2.1 *Geometry model*

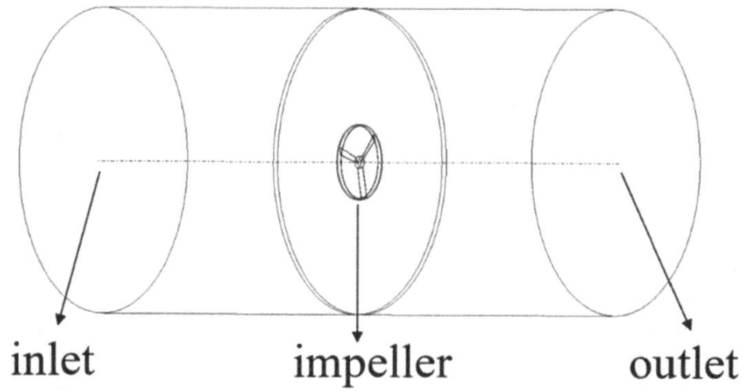

Figure 1. Geometrical model of the axial fan. The left and the right are inlet and outlet, respectively. The one in the middle is the impeller.

The numerical study was carried out on an axial fan (see Figure 1) with a diameter of 1068 mm and three blades on a hub. Driven by the motor, the impeller's rotational speed is set at 290 revolutions per minute. The diameter and length of inlet/outlet are both 4500 mm, as listed in Table 1.

Table 1. Specifications for the axial fan.

Parameter	Unit	Value
Diameter of impeller	mm	1068
Number of blades	–	3
Rotation speed	r/min	290
Inlet/outlet diameter	mm	4500
Inlet/outlet length	mm	4500

2.2 *Governing equations*

The CFD simulations were performed in the commercial software ANSYS Fluent (Fei Zhou, 2022; Lei Zhang, 2014; Shaosong Ni, 2019). In an axial fan, the fluid temperature remains relatively constant, thus it is assumed that the fluid properties are also constant. The flow field within the fan is governed by the steady Navier-Stokes equations. The RNG k-ε model is chosen as the turbulence model. The governing equations include the continuity equation, the momentum equation, and the RNG k-ε equation.

The continuity equation and momentum equation are expressed as:

$$\frac{\partial(\rho u_i)}{\partial x_i} = 0 \tag{1}$$

$$\frac{\partial(\rho u_j u_i)}{\partial x_j} = f_i - \frac{\partial P^*}{\partial x_i} + \frac{\partial\left[\mu_\varepsilon\left(\frac{\partial u_i}{\partial x_j} + \frac{\partial u_j}{\partial x_i}\right)\right]}{\partial x_j} \tag{2}$$

where ρ is the density of the fluid, and t is the time. x_i and x_j are the components of the Cartesian coordinate system in three directions x, y, and z. u_i and u_j are the velocity component u, v, w. P^* is the conversion pressure (including turbulent energy). f_i is the volume force component; μ_ε is the effective viscosity coefficient.

The RNG k-ε model is adopted to consider turbulent flow:

$$\mu_\varepsilon = \mu + \mu_t \tag{3}$$

$$\mu_t = \rho C_\mu \frac{k^2}{\varepsilon} \tag{4}$$

Where μ and μ_t are the molecular viscosity coefficient and the turbulent eddy viscosity coefficient, respectively. The RNG k-ε turbulence model includes turbulence energy and dissipation equations.

2.3 *Meshing and independent verification*

To obtain accurate results, we conducted a grid independence analysis to evaluate the impact of grid size on numerical results. The tetrahedral mesh is used for the entire computational domain of the axial fan. The grid is well dense in the impeller region and coarse in the inlet/outlet region. The resolution of the grid has been tested to ensure its independence. Five sets of grids were selected, as follows: 1.26 million, 2.6 million, 4.06 million, 7.1 million, and 16.27 million. The results are shown in Figure 2. When the number of grids exceeds 2.6 million, there is no significant change in the flow rate at the outlet. Taking computational resources into account, the entire computational model opted for 4.06 million grids.

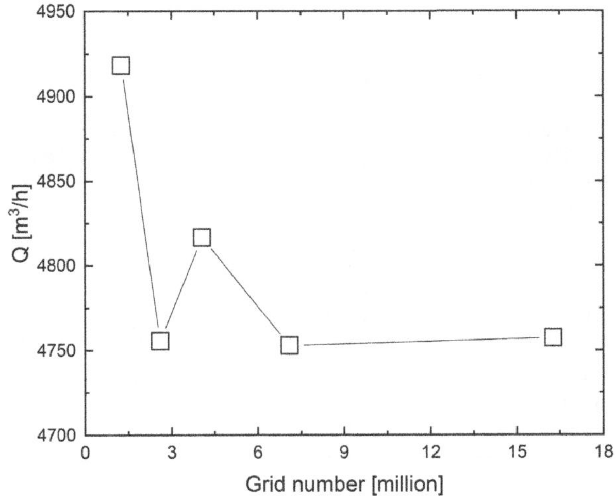

Figure 2. Validation of flow rate (Q) with the number of grids.

3 RESULTS AND DISCUSSION

A preliminary test was conducted on the performance of an axial flow fan, utilizing 4.06 million grids. The test was performed with a density of 1.225 kg/m^3, a viscosity of 1.79×10^{-5} $Pa \cdot s$, a standard atmospheric pressure of 101325 Pa, and a temperature of 15°C. Here we study how the temperature impact on the performance of the axial fan. In the cases of low temperature, the temperature is changed from −40°C and 200°C, as shown in Table 2. For examples, The extreme minimum temperature is −53°C at Mohe City during winter, the operating temperature of computer cooler varies from 20°C to 90°C, and the operating temperature of the air fryer changes from 20°C to 200°C. Table 2 shows the variation of dry air density and viscosity with temperature. As the temperature increases from −40°C to 200°C, the density of dry air decreases to 49% of its initial value, which means that dry air becomes lighter at higher temperatures. As the temperature rises from −40°C to 200°C, the viscosity of dry air increases to 171% of its initial value, which means that dry air becomes more viscous at higher temperatures. In further, Table 3 shows the cases of high temperature from −50°C and 1200°C. One example is that the operating temperature of an aeroengine can reach up to 2000°C. In Table 3, as the temperature rises from −50°C to 1200°C, the density of dry air decreases to 15% of its initial value. As the temperature rises from −50°C to 1200°C, the viscosity of dry air increases to 366% of its initial value.

Table 2. Physical properties of dry air.

Temperature (T/°C)	Density ($\rho/(kg/m^3)$)	Viscosity ($\mu/(10^{-5} Pa \cdot s)$)
−40	1.515	1.52
−20	1.395	1.62
0	1.293	1.72
20	1.205	1.81
40	1.128	1.91
60	1.06	2.01
80	1	2.11
120	0.898	2.28
160	0.815	2.45
200	0.746	2.6

Table 3. Physical properties of dry air.

Temperature (T/°C)	Density ($\rho/(kg/m^3)$)	Viscosity ($\mu/(10^{-5} Pa \cdot s)$)
−50	1.584	1.46
0	1.293	1.72
100	0.946	2.19
200	0.746	2.6
400	0.524	3.3
600	0.404	3.91
800	0.329	4.43
1000	0.277	4.9
1200	0.239	5.35

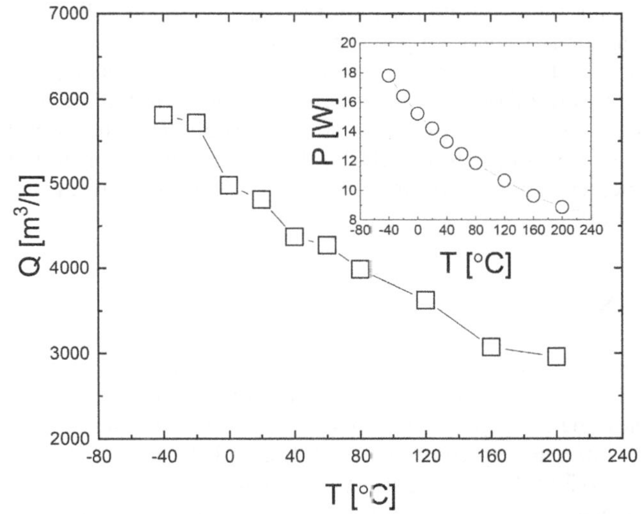

Figure 3. The evolution of flow rate (Q) and power (P) with temperature (T) in the cases of low temperature.

Figure 3 shows the evolution of flow rate (Q) and power (P) of impeller with temperature (T) in the cases of low temperature, in which the power is the required power for the impeller as shown in the insert. In the whole, the flow rate of outlet and the power decrease as the temperature increases. The flow rate of outlet decreases to 51% of its initial value and the power of impeller decreases to 50% of its initial value. One example is the computer cooling fan that the corresponding flow rate of outlet decreases to the initial 81% as the computer cooling fan heats up from 20°C to 90°C. Another example is the air fryer that the corresponding flow rate of outlet decreases to the initial 61% as the air fryer fan heats up from 20°C to 200°C. These two examples indicate that temperature plays a key role in axial fans, which means that the design and numerical simulation of fans must take into account the impact of temperature.

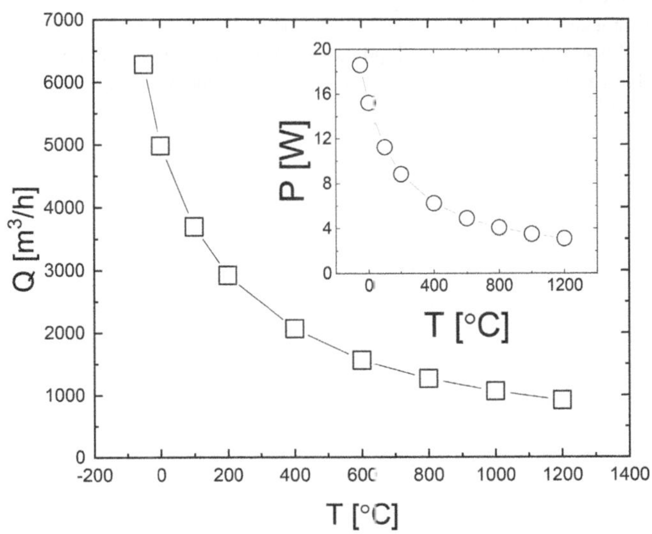

Figure 4. The evolution of flow rate (Q) and power (P) with temperature (T) in the cases of high temperature.

Figure 4 represents the evolution of flow rate (Q) and power (P) of impeller with temperature (T) in the cases of high temperature. In the whole, the flow rate of outlet and the power decrease as the temperature increases, and the slope decreases as the temperature increases. The flow rate of outlet decreases to 15% of its initial value and the power of impeller decreases to 17% of its initial value. One example is the aircraft engine. As the aircraft engine heats up from 20°C to 1200°C, the corresponding air output decreases to the initial 19%. It indicates that temperature plays a key role in axial fans, which means the same conclusion of Figure 3.

4 CONCLUSIONS

In this paper, we investigate the impact of temperature on the performance of an axial fan by ANSYS Fluent. The numerical results show that temperature plays a key role in the axial fans. For examples, the corresponding flow rate of outlet decreases to the initial 81% as the computer cooling fan heats up from 20°C to 90°C, and the initial 61% as the air fryer fan heats up from 20°C to 200°C, and the corresponding air output decreases to the initial 19% as the aircraft engine heats up from 20°C to 1200°C. Those results show that temperature plays a crucial role in the axial fans, and the design and numerical simulation of fans must take into account the impact of temperature.

REFERENCES

Antoine Moreau, Sebastien Guerin, Lars Enghard. (2011). Axial-Flow ventilation fan design through multi-objective optimization to enhance aerodynamic performance. *Journal of Fluids Engineering*, 133.

Fei Zhou, Yu Zhou and Xuemin Ye. (2022). Influence of a winglet combined with a groove tip on the performances of a variable-pitch axial flow fan. *FDMP*, 18, 4.

Lei Zhang, Jinhua Lang, Kuan Jiang and Songling Wang. (2014). Simulation of entropy generation under stall conditions in a centrifugal fan. *Entropy*, 16, 3573-3589.

Sarraf C., Nouri H., Ravelet F., Bakir F. (2011). Experimental study of blade thickness effects on the overall and local performances of a Controlled Vortex Designed axial-flow fan. *Experimental Thermal and Fluid Science*, 35, 684-693.

Shaosong Ni, Wenbin Cao, Jun Xu, Yingdong Wang and Wei Zhang. (2019). Effects of an inclined blade on the performance of a sirocco fan. *Applied Science*, 9, 3154.

Zivan Spasic, Milos Jovanovic and Jasmina Bogdanovic-Jovanovic. (2018). Design and performance of low-pressure reversible axial fan with doubly curved profiles of blades. *Journal of Mechanical Science and Technology*, 32, 3707-3712.

Fluid Mechanics and Industrial Applications – Fang (Ed.)

Multi-scenario calibration method for double-Gaussian wake model based on DE-SQP cooperative optimization

Na Ren, Yuhao Li and Xiangmeng Xia
Mingyang Smart Energy, Control Algorithm Room at Sea, Zhongshan, Guangdong, China

ABSTRACT: Against the backdrop of global energy transition, the issues of power generation loss and equipment fatigue caused by wind farm wake effects are becoming increasingly prominent. This paper addresses the limited generalization capability of existing double-Gaussian wake models by proposing a Differential Evolution-Sequential Quadratic Programming (DE-SQP) cooperative optimization framework. This framework enhances multi-scenario prediction accuracy through the calibration of 15 empirical parameters within the model. Firstly, wake datasets under three atmospheric stability conditions – stable, neutral, and convective – are generated based on CFD simulations. Secondly, using the root mean square error (RMSE) of wind turbine power prediction in the wake region as the objective function, the DE algorithm is employed for global search of the parameter space, followed by local fine-tuning using the SQP algorithm. Finally, comparative validation demonstrates that the calibrated model reduces power prediction errors by 96.97%–99.21%, with the maximum error not exceeding 50.75 kW. This method significantly improves the robustness of the double-Gaussian model across the entire wake region, providing an efficient analytical tool for intelligent wind farm control.

1 INTRODUCTION

Driven by global energy transition and carbon neutrality goals, wind energy, as a core clean energy source, is developing rapidly. According to statistics from the International Energy Agency (IEA), as of the first half of 2025, the global cumulative installed wind power capacity has exceeded 1,136 GW. It is projected that from 2025 to 2030, the global new installed capacity for both onshore and offshore wind power will reach 982 GW, and the proportion of wind power in the power system will continue to rise (GWEC Global Wind Report 2025). However, as wind farms scale up and turbine density increases, the problems of power generation loss and equipment fatigue caused by wake effects are becoming increasingly prominent. Studies show that annual power generation losses due to wake interference in typical wind farms can reach 5%–10% (Zhao L Y 2022), and these losses may further increase under complex terrain and atmospheric stability conditions (Chen Y, 2025) (MARTIN KüHN, 2019). Additionally, the additional turbulence generated by wake effects exacerbates fatigue damage to components such as blades, rotors, and towers (Kang Y, 2024; Yao T C, 2023; Zhang J, 2024), increasing operation and maintenance costs. Therefore, accurate wake prediction and control are crucial for enhancing wind farm operational efficiency and equipment safety.

Wind Farm Flow Control (WFFC) serves as a core strategy to mitigate wake effects. By adjusting turbine operating states (such as pitch, yaw, and torque control) to actively optimize flow field distribution, it aims to achieve goals such as total farm power maximization, load optimization, and power tracking. Existing control strategies can be categorized into two types: one involves adjusting the yaw angle of wind turbines to laterally shift the wake position, known as wake steering optimization (Nakhchi M E, 2022); the other involves adjusting the thrust coefficient to directly reduce the velocity deficit downstream, termed axial induction

DOI: 10.1201/9781003679561-7

optimization (BARTL J, 2016). Research indicates that the more severe the wake effects within a wind farm, the more significant the power gains achievable through these optimization strategies (Liu Y G, 2024). Heck *et al.* (2025) and Li *et al.* (2022) both validated the effectiveness of these two control strategies using Large Eddy Simulation (LES) and achieved wake control for specific wind farms. Although numerical simulation methods like LES can accurately reproduce wake field characteristics, their high computational cost makes them impractical for real-time control needs. Therefore, engineering applications urgently require efficient and reliable analytical wake models. Early models (such as the Jensen model and Gaussian models), based on simplified momentum theory, were somewhat effective for predicting velocity deficits in the far wake region under uniform atmospheric conditions. However, the limitations of their theoretical assumptions resulted in insufficient predictive capability in the near wake region. In recent years, CFD simulation technology based on the Actuator Line Model (ALM) has provided new avenues for model improvement. Scholars like Qian *et al.* (2025) discovered through ALM-CFD coupled simulations that atmospheric turbulence intensity and turbine thrust coefficient are key parameters governing wake velocity recovery and near-wake length. Based on this, they proposed the double-Gaussian wake model. This model uses two Gaussian functions to describe the velocity deficit distribution in the near and far wake regions respectively, significantly improving prediction accuracy in the near wake. Wang *et al.* (2024) introduced Symbolic Regression (SR) genetic algorithms to develop a double-Gaussian wake model with explicit mathematical expressions, enabling stable and accurate prediction of velocity loss at any location within the entire wake region.

However, most existing double-Gaussian models are calibrated based on numerical simulation results for specific wind turbine models under specific atmospheric environments, exhibiting significant limitations in generalization capability. To address these issues, this paper proposes a joint optimization framework based on Differential Evolution (DE) and Sequential Quadratic Programming (SQP) for multi-scenario parameter calibration of the double-Gaussian model. The specific research contents are as follows: 1) Utilize CFD simulations to generate wake datasets under three typical atmospheric stability conditions: stable, neutral, and convective; 2) Employ the prediction error of wind turbine power in the wake region as the objective function, use the DE algorithm for global search of model parameters, and combine it with the SQP algorithm for local fine-tuning of the DE results to enhance parameter accuracy; 3) Conduct comparative validation to assess the calibrated model's accuracy in predicting wake power under different atmospheric stability conditions.

2 MODEL AND OPTIMIZATION METHOD

2.1 *DE-SQP joint optimization framework*

To enhance the prediction accuracy of the double-Gaussian wake model under different atmospheric stability conditions, this paper proposes a joint optimization framework based on Differential Evolution (DE) and Sequential Quadratic Programming (SQP). This framework achieves accurate prediction of wind turbine performance in the wake region by optimizing the empirical parameters within the model. The analytical wake model adopted in this study (QIAN G W, 2025) includes 3 input parameters (thrust coefficient, wind speed, turbulence intensity) and 15 empirical parameters directly related to wake characteristics.

2.1.1 *Differential evolution algorithm*

The Differential Evolution algorithm (DE) is a population-based global optimization method suitable for optimization problems in multimodal, high-dimensional parameter spaces. For the 15 empirical parameters of the double-Gaussian model, they are denoted as a parameter vector $\mathbf{x} \in \mathbb{R}^{15}$. The core procedure of the DE algorithm is as follows:

First, an initial population consisting of NP individuals is generated. Each individual $\mathbf{x}_i (i = 1, 2, \ldots, NP)$ corresponds to a set of parameter vectors. During the mutation operation phase, for a target individual $\mathbf{x}_i^{(g)}$ in the g-th generation population, a mutation vector is generated using a differential strategy:

$$\mathbf{v}_i^{(g+1)} = \mathbf{x}_{r1}^{(g)} + F \cdot \left(\mathbf{x}_{r2}^{(g)} - \mathbf{x}_{r3}^{(g)}\right) \tag{1}$$

where $r1$, $r2$, $r3$ are three mutually distinct individual indices randomly selected from the population, and $F \in [0, 1]$ is the mutation factor controlling the differential magnitude.The crossover operation mixes the mutation vector $\mathbf{v}_i^{(g+1)}$ with the target individual $\mathbf{x}_i^{(g)}$ dimension-wise to generate the trial vector $\mathbf{u}_i^{(g+1)}$. For the j-th dimensional component:

$$u_{ij}^{(g+1)} = \begin{cases} v_{ij}^{(g+1)}, & \text{if } \text{rand}_j \le CR \text{ or } j = j_{rand} \\ x_{ij}^{(g)}, & \text{others} \end{cases} \tag{2}$$

where $CR \in [0, 1]$ is the crossover probability, and j_{rand} is a randomly selected dimension index. During the selection operation, the fitness of the trial vector $\mathbf{u}_i^{(g+1)}$ is compared with that of the target individual $\mathbf{x}_i^{(g)}$, and the superior individual is retained for the next generation population. The fitness function is defined as the root mean square error (RMSE) between the model predicted power and the LES simulated power:

$$\text{Fitness}(\mathbf{x}) = \sqrt{\frac{1}{N}\sum_{i=1}^{N} \left(P_{\text{LES},i} - P_{\text{model,i}}(\mathbf{x})\right)^2} \tag{3}$$

where $P_{\text{LES},i}$ is the average power of the wind turbine at the i-th monitoring section in the CFD simulation, $P_{\text{model},i}(\mathbf{x})$ is the power predicted by the double-Gaussian model at the same location based on parameters $\mathbf{x}$, and N is the total number of monitoring sections. The algorithm terminates when the maximum number of iterations is reached or the fitness change falls below a threshold, outputting the current optimal parameter vector $\mathbf{x}_{DE}$.

2.1.2 *Sequential quadratic programming algorithm*

After DE completes the global search, the SQP algorithm performs local refinement optimization using $\mathbf{x}_{DE}$ as the initial point. The core steps are as follows:

At the k-th iteration, a QP subproblem is constructed based on current parameters $\mathbf{x}_k$:

$$\begin{aligned} &\min_{\mathbf{d}} \frac{1}{2}\mathbf{d}^T \mathbf{H}_k \mathbf{d} - \nabla f(\mathbf{x}_k)^T \mathbf{d} \\ &\text{s.t. } \nabla g_i(\mathbf{x}_k)^T \mathbf{d} + g_i(\mathbf{x}_k) \le 0, \quad i = 1, \ldots, m \end{aligned} \tag{4}$$

Where $\mathbf{d}$ is the search direction vector, $\mathbf{H}_k$ is the BFGS approximation of the Hessian matrix of the objective function, $g_i(\mathbf{x}_k)$ represents the parameter constraints. After solving this subproblem to obtain the optimal search direction $\mathbf{d}_k$, the step size is determined through one-dimensional line search:

$$\alpha_k = \arg \min f(\mathbf{x}_k + \alpha \mathbf{d}_k) \tag{5}$$

Update the parameter vector: $\mathbf{x}_{k+1} = \mathbf{x}_k + \alpha_k \mathbf{d}_k$. The iteration process terminates when the first-order Karush-Kuhn-Tucker (KKT) necessary conditions are satisfied:

$$\begin{aligned} &\nabla f(\mathbf{x}^*) + \sum \mu_i \nabla g_i(\mathbf{x}^*) = \mathbf{0} \\ &\mu_i g_i(\mathbf{x}^*) = 0,\ \mu_i \ge 0,\ g_i(\mathbf{x}^*) \le 0 \end{aligned} \tag{6}$$

Where μ_i is the Lagrange multiplier corresponding to the inequality constraint, $\mathbf{x}^*$ is the locally optimal solution. Through this process, model parameters that satisfy physical constraints and minimize prediction error are obtained.

2.2 *CFD numerical modeling based on ALM*

2.2.1 *Actuator line model*

The Actuator Line Model (ALM) is employed to model wind turbines. Compared to the Actuator Disk Model (NADERI S, 2018), ALM discretizes each blade into a series of elements. The aerodynamic forces on each element are computed based on local inflow conditions and airfoil data:

$$\mathbf{F}_i = \mathbf{L}_i + \mathbf{D}_i \tag{7}$$

Where $\mathbf{L}_i$ and $\mathbf{D}_i$ represent the lift and drag forces on element i, respectively. Their expressions are defined as follows:

$$\begin{aligned} \mathbf{L}_i &= \frac{1}{2}\rho \mathbf{u}_i^2 c_i t_i \Delta s_i C_{l,i} \\ \mathbf{D}_i &= \frac{1}{2}\rho \mathbf{u}_i^2 c_i t_i \Delta s_i C_{d,i} \end{aligned} \tag{8}$$

Where ρ is the free-stream air density, $\mathbf{u}_i$ is the local inflow velocity at element i, $C_{l,i}$ and $C_{d,i}$ are the lift and drag coefficients of airfoil i at angle of attack α, respectively, Δs_i is the element length, c_i and t_i are the chord length and thickness of the airfoil.

To avoid numerical instability, the force $\mathbf{F}_i$ is not applied directly at discrete points. Instead, it is distributed to surrounding grids using a smoothing function (DIAZ G P N, 2023):

$$\eta_\epsilon(\mathbf{r}) = \frac{1}{\epsilon^3 \pi^{\frac{3}{2}}} \exp\left(-\frac{|\mathbf{r}|^2}{\epsilon^2}\right) \tag{9}$$

where $\mathbf{r}$ is the distance vector from the grid point to the actuator line element, and ϵ denotes the smoothing radius (with $\epsilon = 2\Delta x$, where Δx is the grid size). The volume force at any spatial point $\mathbf{x}$ can then be expressed as:

$$\mathbf{F}(\mathrm{x}) = \sum\nolimits_{i=1}^{N} \mathbf{F}_i \cdot \eta_\epsilon(\mathbf{x} - \mathbf{x}_i) \tag{10}$$

where N denotes the total number of blade elements, and $\mathbf{x}_i$ represents the position vector of element i.

This study employs an onshore wind turbine as the research object, with a hub height of 109 m and a rotor diameter of 178 m. The variations of its thrust coefficient and power curve with wind speed are illustrated in Figure 1.

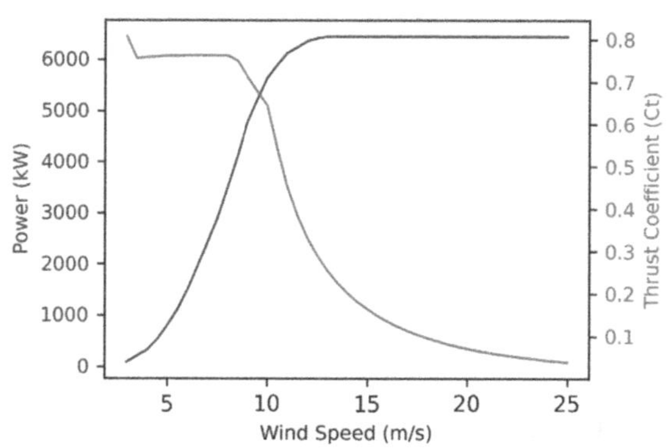

Figure 1. Power curve and thrust coefficient vs. wind speed.

2.2.2 *Inlet turbulence generation*

To establish inlet turbulence boundary conditions under different atmospheric stabilities, this study adopts the precursor simulation method (MELAKU A F, 2021) (PHUC P V, 2018). This approach generates inflow data conforming to realistic atmospheric characteristics through long-duration turbulence field evolution. The total simulation duration is set to 32,400 s, with the initial 28,800 s dedicated to developing a fully developed turbulence field. Data from the subsequent 3,600 s are extracted to serve as inlet boundary conditions for downstream wind turbine wake simulations.

Atmospheric stability is characterized by the Monin-Obukhov length L (KARL-HEINZ, 1995), defined as:

$$L = -\frac{u_\tau^3 \theta_0}{\kappa g q} \tag{11}$$

where κ denotes the von Kármán constant, g is the gravitational acceleration (units: m/s^2), q represents the surface heat flux (units: $K \cdot m/s$), and θ_0 is the reference temperature (units: K). When $L < 0$, buoyancy enhances turbulence, corresponding to a convective atmospheric boundary layer; when $L > 0$, buoyancy suppresses turbulence, characteristic of a stable atmospheric boundary layer; and as $L \to \infty$, conditions approach a neutral atmospheric boundary layer.

Near-surface boundary conditions are simulated using a wall model based on the Monin-Obukhov Similarity Theory (WANG Z W, 2025) (VAN DER, 2017). This theory correlates turbulent transport processes with local conditions:

$$u(z) = \frac{u_\tau}{\kappa}\left[\ln\left(\frac{z}{z_0}\right) - \psi_m\left(\frac{z}{L}\right) + \psi_m\left(\frac{z_0}{L}\right)\right] \tag{12}$$

$$\theta(z) = \theta(z_0) + \frac{\theta_\tau}{\kappa}\left[\ln\left(\frac{z}{z_0}\right) - \psi_h\left(\frac{z}{L}\right) + \psi_h\left(\frac{z_0}{L}\right)\right] \tag{13}$$

Where u_τ is the friction velocity, z_0 denotes the roughness length, θ_τ represents the characteristic temperature, ψ_m and ψ_h are the stability functions for momentum and heat transfer, respectively.

For stable atmospheric boundary conditions:

$$\psi_m = -\gamma_m \frac{z}{L}, \ \psi_h = -\gamma_h \frac{z}{L} \tag{14}$$

where $\gamma_m = 4.8$ and $\gamma_h = 7.8$.

For convective atmospheric boundary conditions:

$$\begin{aligned} \psi_m &= 2\ln\left(\frac{1+x}{2}\right) + \ln\left(\frac{1+x^2}{2}\right) - 2\arctan x + \frac{\pi}{2}, x = \left(1 - \beta_m \tfrac{z_b}{L}\right)^{\frac{1}{4}} \\ \psi_h &= \ln\left(\frac{1+y}{2}\right), y = \left(1 - \beta_h \tfrac{z_b}{L}\right)^{\frac{1}{2}} \end{aligned} \tag{15}$$

Where, $\beta_m = 16$ and $\beta_h = 16$.

The fundamental simulation parameters for all cases were set as follows: reference temperature $\theta_0 = 277$ K, inflow velocity $U_\infty = 10$ m/s, roughness length $z_0 = 0.001$ m, computational domain horizontal dimensions 3520 m × 2560 m, and base grid size 20 m × 20 m. The simulation domain height and wall boundary condition parameters were adjusted according to the atmospheric stability. Specific parameter settings (HODGSON E L, 2023) are detailed in Table 1:

Table 1. Parameter settings of different atmospheric stability generation mechanisms.

Case	Initial Temp (K)	Lower BC	Height (m)
CBL	277	–	800
SBL	$z \leq 100m, 277$ $z > 100m, 277 + 0.01z$	Cooling rate $= 0.25K/h$	1600
NBL	$z \leq 937m, 277$ $937 - 1063m, 277 + 0.0635z$ $z > 1063m, 277 + 0.003z$	Surface heat flux $= 0.24K \cdot m \cdot s^{-1}$	2000

Table 2. Simulation boundary condition setting.

Boundary	x-low	x- high	y-low	y-high	z-low	z- high
Type	mass-inflow	pressure-out	slip	slip	no-slip	slip

3 RESULTS ANALYSIS

3.1 *Inflow turbulence and wake characteristics analysis*

Vertical profiles of average wind speed, temperature, and turbulent kinetic energy (TKE) under different atmospheric stability conditions are presented in Figure 2. Under convective conditions, surface heating induces thermal buoyancy effects that continuously transport high-TKE airflow from near-surface levels to higher altitudes, driving upward boundary layer expansion and resulting in the maximum boundary layer thickness of approximately 1300 m. Stable conditions feature surface cooling where buoyancy suppresses turbulence, inhibiting vertical diffusion and yielding the minimum boundary layer thickness of about 350 m. Neutral conditions exhibit minimal temperature variation with height, with an intermediate boundary layer thickness of approximately 800 m. Analysis of TKE vertical distribution reveals that under stable conditions, the near-surface maximum TKE reaches 0.181 m^2s^{-2}, decaying linearly with height at a gradient of 0.804 m^2s^{-2}/km. Neutral conditions show a near-surface maximum TKE of 0.136 m^2s^{-2} decaying approximately logarithmically. Convective conditions generate the highest near-surface TKE (0.229 m^2s^{-2}), with TKE decay displaying distinct vertical staging: logarithmic distribution between 100-1000 m (slower decay) transitioning to linear decay between 1000-1300 m (accelerated decay). These patterns confirm that precursor-simulated turbulence fields adhere to atmospheric boundary layer physics, establishing reliable boundary conditions for wind turbine wake analysis.

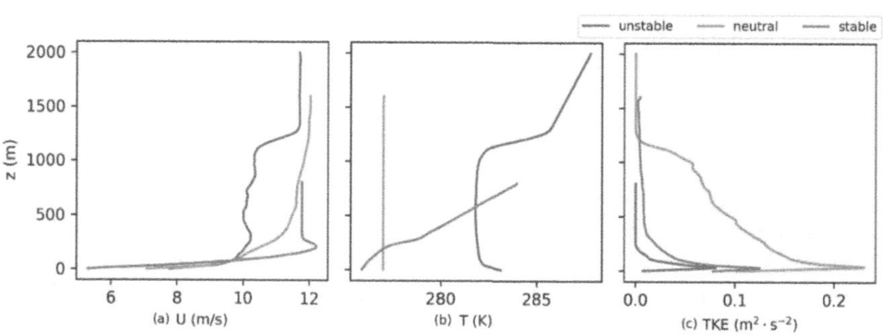

Figure 2. Average velocity profile, temperature profile and turbulent kinetic energy profile at the inlet under different atmospheric stability.

Figure 3 illustrates the time-averaged velocity field distribution on the hub-height horizontal plane under three typical atmospheric stability conditions. The simulation results reveal that the downstream influence distance of the wake is shortest under convective conditions, intermediate under neutral conditions, and longest under stable conditions. Further analysis of the dimensionless velocity deficit at different downstream distances (Figure 4) indicates that the peak velocity deficit in the near-wake region (1D-3D) is similar across conditions, with the primary differences manifesting in the far-wake region (5D-15D). To assess the wake's impact on downstream turbine power generation performance, six monitoring Sections (3D, 5D, 7D, 9D, 11D, 13D) were arranged on the hub-height horizontal plane. Eighteen equally spaced sampling points spanning a 1D range were set along each section's lateral direction. The theoretical power curve of the turbine was calculated based on the average velocity at each sampling point, as shown in Figure 5. The results demonstrate that the power output differences at the 3D location under the three stability conditions are relatively small, with the maximum power discrepancy not exceeding 300 kW. In contrast, significant power differences emerge in the far-wake region beyond 5D. Specifically, during the 5D to 9D far-wake stage, the power difference under varying atmospheric conditions can reach 1000 kW. Although this power difference subsequently decreases with increasing distance, it remains substantially larger than that observed in the near-wake stage.

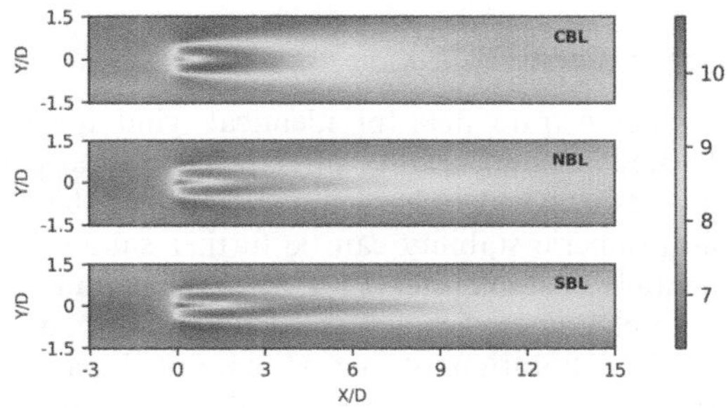

Figure 3. Time-averaged velocity field distribution of hub height horizontal plane under three typical atmospheric stability.

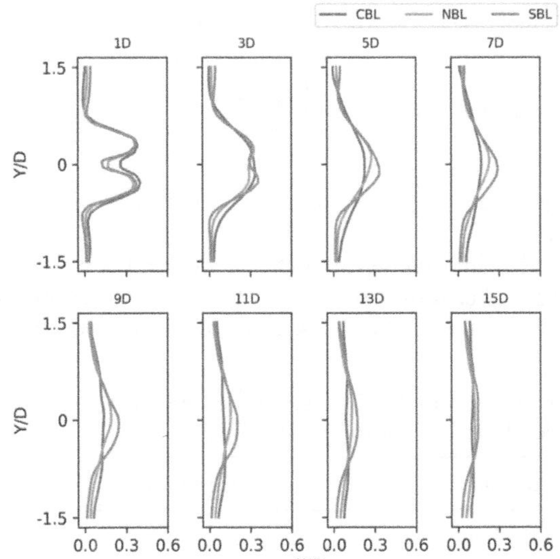

Figure 4. Dimensionless velocity loss at different downstream distances.

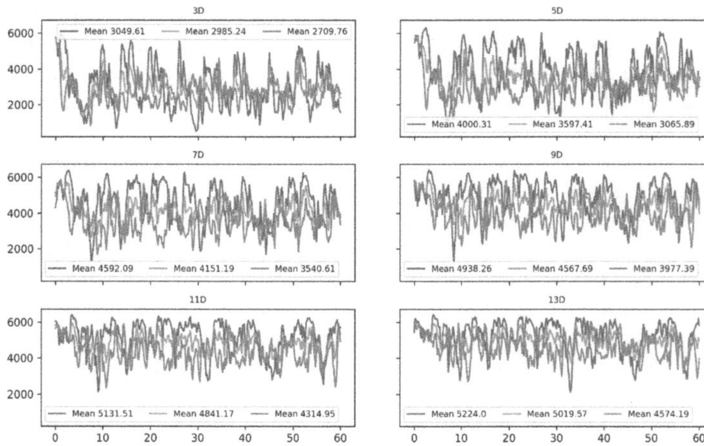

Figure 5. The variation of wind turbine power with downstream distance under different atmospheric stability.

3.2 *Model calibration effectiveness verification*

CFD simulation results demonstrate that for identical wind turbine models, even with similar horizontal wind speeds at hub height, differences in atmospheric boundary conditions lead to significantly distinct wake characteristics behind the rotor. In actual atmospheric environments, atmospheric stability can be further subdivided into weakly stable, near-neutral stable, moderately unstable, etc. (LI Y Z, 2025), making it highly challenging to numerically replicate wake characteristics across all boundary conditions. Therefore, this section conducts systematic calibration of the 15 empirical parameters in the double-Gaussian wake model, utilizing the Differential Evolution-Sequential Quadratic Programming (DE-SQP) co-optimization framework proposed in Section 2.1.

The minimization of the root mean square error (RMSE) between the model-predicted turbine power (P_{model}) and the LES-simulated power (P_{LES}) at the 3D, 7D, and 11D sections within the wake region serves as the optimization objective. This objective function is also the core metric for subsequent model accuracy evaluation:

$$\min \text{RMSE}(\mathbf{x}) = \sqrt{\frac{1}{N}\sum_{i=1}^{N}\left(P_{\text{LES},i} - P_{\text{model,i}}(\mathbf{x})\right)^2} \tag{16}$$

Here, $N = 3$ represents the number of monitoring sections, and $\mathbf{x}$ denotes the parameter vector to be optimized.

During the DE global search phase, the initial population size was set to $NP = 15$, with a mutation factor $F = 0.5$ and crossover probability $CR = 0.7$. After 30 iterations, the initial solution $\mathbf{x}_{DE}$ was obtained. Subsequently, SQP-based local refinement was performed using $\mathbf{x}_{DE}$ as the

Table 3. Comparison of RMSE of power prediction before and after model correction under different atmospheric stability.

Case	default (kW)	optimized (kW)	reduction rate (%)
CBL	1094.59	13.62	98.76
NBL	1091.81	8.58	99.21
SBL	851.15	25.76	96.97

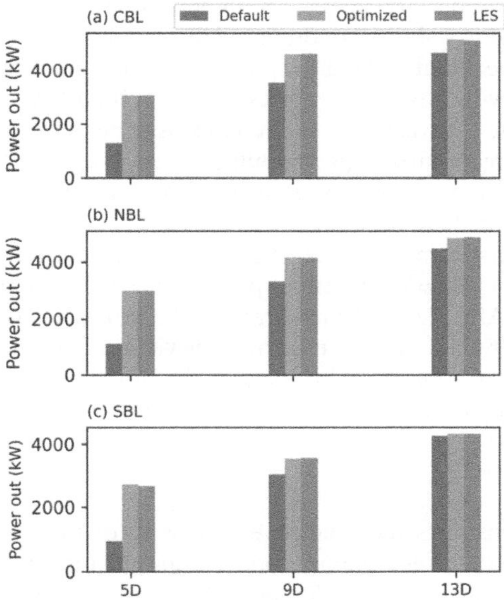

Figure 6. Comparison of power prediction values and LES results before and after model correction at 5D, 9D and 13D positions under different atmospheric stability.

starting point. The iterative solving process adhered to the KKT condition constraints (Equation (6)), with a convergence threshold of 10^{-5}. As shown in Table 3, the calibrated model exhibits a substantial reduction in power prediction RMSE across all atmospheric stability conditions.

A core value of analytical wake models lies in their ability to efficiently and accurately predict the impact of wakes on downstream wind turbine power generation efficiency, providing crucial support for real-time online wake control. Therefore, a further evaluation of the calibrated model's predictive capability is warranted, estimating the average power of turbines at different downstream distances (5D, 9D, 13D). The results, shown in Figure 6, reveal that the prediction error of the uncorrected double-Gaussian model relative to LES results exhibits an inverse relationship with downstream distance: error increases with proximity. The maximum error occurs under neutral conditions at the 3D section, while the minimum prediction error occurs under stable conditions at the 13D section (see blue and green bars in Figure 6). This phenomenon aligns with the characteristic limitation of current mainstream analytical models, which typically perform better in the far-wake region but lack precision in the near-wake region. This is likely attributable to the near-wake flow field being significantly influenced by complex tip and root vortices, whereas the far-wake flow field is predominantly governed by atmospheric boundary conditions, highlighting the aerodynamic characteristics of the turbine as a key factor influencing near-wake features. The calibrated analytical model effectively mitigates this issue, significantly enhancing prediction accuracy across the entire wake region. As illustrated by the orange and green bars in Figure 6, the maximum discrepancy in predicted average power across the three atmospheric stability conditions is 75 kW, occurring under neutral conditions at 5D. The minimum power prediction difference is 0.76 kW, observed under convective conditions at 7D.

4 CONCLUSIONS AND OUTLOOK

This study systematically calibrated the empirical parameters of a double-Gaussian analytical wake model using a Differential Evolution-Sequential Quadratic Programming (DE-

SQP) co-optimization framework, based on numerical simulations of wind turbine wakes under varying atmospheric stability conditions. The key findings are summarized as follows:

CFD simulations reveal that wake diffusion characteristics are significantly influenced by atmospheric stability, with power differences of up to 1000 kW observed across different atmospheric conditions. Convective conditions exhibit the shortest wake impact distance and the highest downstream turbine power output. Conversely, stable conditions feature the longest wake impact distance, resulting in exacerbated power attenuation. Neutral conditions demonstrate intermediate behavior between these two extremes.

Consistent with mainstream analytical models, the uncorrected double-Gaussian model under default parameters shows substantial prediction errors in the near-wake region compared to LES results. After DE-SQP calibration, the model's power prediction errors are reduced by 96.97% to 99.21%, with a maximum deviation from LES results not exceeding 50.75 kW. This validates both the robustness of the double-Gaussian model in wake prediction and the efficiency of the co-optimization framework.

4.1 *Outlook*

While this study demonstrates the predictive capability of the double-Gaussian analytical wake model through parameter optimization, the current model does not account for the influence of wind shear on wake development, limiting its accuracy in the vertical plane. Future work will focus on developing a 3D double-Gaussian wake model incorporating wind shear effects to enhance predictive performance in complex terrain. Additionally, empirical parameters for sub-models (e.g., wake deflection and velocity superposition) will be further optimized using CFD simulations and field measurement data. These advancements aim to provide robust technical support for wake management in large-scale wind farms.

REFERENCES

Bartl J, Sætran L. (2016). Experimental testing of axial induction based control strategies for wake control and wind farm optimization; proceedings of the Conference on Science of Making Torque from Wind (TORQUE), Munich, Germany, F Oct 05-07, [C]. 2016.

Chen Y, Yan B W, Yu M, *et al.* (2025). Wind tunnel study of wind turbine wake characteristics over two-dimensional hill considering the effects of terrain slope and turbine position [J]. *Applied Energy*, 380.

Diaz G P N, otero A D, Asmuth H, *et al.* (2023). Actuator line model using simplified force calculation methods [J]. *Wind Energy Science*, 8(3): 363-82.

GWEC Global Wind Report 2025 [R], (2025).

Heck K S, Liew J, Upfal I M L, *et al.* (2023). Joint yaw-induction control optimization forwind farms [J]. *Wind Energy Science*, 2025, CC B Y 4.0.

Hodgson E L, Souaiby M, Troldborg N, *et al.* (2023). Cross-code verification of non-neutral ABL and single wind turbine wake modelling in LES; proceedings of the 8th Wake Conference, Visby, SWEDEN, F Jun 20-22, 2023 [C].

Kang Y, Lee S. (2024). Wake effect on floating offshore wind turbine fatigue load [J]. *Physics of Fluids*, 36(12).

Karl-Heinz Bernhardt B. (1995). On the interpretation of the Monin-Obukhov's length [J]. *Meteor-ologische Zeitschrift* , N.F.4: 81-2.

Li B L, He J, Ge M W, *et al.* (2022). Study of three wake control strategies for power maximization of offshore wind farms with different layouts [J]. *Energy Conversion and Management*, 268.

Liu Y G, Zhao Z Z, Liu Y, *et al.* (2024). Combined wake control of aligned wind turbines for power optimization based on a 3D wake model considering secondary wake steering [J]. *Energy*, 308.

Li Y Z, Okaze T, Ikegaya N. (2025). Turbulent characteristics in the atmospheric boundary layer under neutral and unstable thermal stratifications based on LES [J]. *J Wind Eng Ind Aerod*, 263.

Martin KÜHN S R. (2019). Influence of atmospheric stability on wind farm control [J].

Melaku A F, Bitsuamlak G T. (2021). A divergence-free inflow turbulence generator using spectral representation method for large-eddy simulation of ABL flows [J]. *Journal of Wind Engineering and Industrial Aerodynamics*, 212.

Naderi S, Parvanehmasiha S, Torabi F. (2018). Modeling of horizontal axis wind turbine wakes in Horns Rev offshore wind farm using an improved actuator disc model coupled with computational fluid dynamic [J]. *Energy Conversion and Management*, 171: 953-68.

Nakhchi M E, Naung S W, Rahmati M. (2022). A novel hybrid control strategy of wind turbine wakes in tandem configuration to improve power production [J]. *Energy Conversion and Management*, 260.

Phuc P V, Nozu T, Kikuchi H, *et al.* (2018). Wind pressure distributions on buildings using the coher-ent structure smagorinsky model for LES [J]. *Computation*, 6(2).

Qian G W, Ishihara T. (2025). A novel double-Gaussian full wake model for wind turbines considering dependence on thrust coefficient and ambient turbulence intensity [J]. *Applied Energy*, 391.

van der Laan M P, Kelly M C, Sorensen N N. (2017). A new k-epsilon model consistent with Mon-in-Obukhov similarity theory [J]. *Wind Energy*, 20(3): 479-89.

Wang D, Chen Y T, Chen S Y. (2024). Discovering an interpretable mathematical expression for a full wind-turbine wake with artificial intelligence enhanced symbolic regression [J]. *Physics of Fluids*, 36(10).

Wang Z W, Jacob J, Marlow F, *et al.* (2025). Lattice Boltzmann method-based large eddy simulation of the influence of atmospheric stability on wind farm wake [J]. *Physics of Fluids*, 37(6).

Yao T C, Lu Q, Wang Y P, *et al.* (2023). Numerical investigation of wake-induced lifetime fatigue load of two floating wind turbines in tandem with different spacings [J]. *Ocean Engineering*, 285.

Zhao L Y, Xue L, Li Z Q, *et al.* (2022). Progress on offshore wind farm dynamic wake management for energy [J]. *Journal of Marine Science and Engineering*, 10(10).

Zhang J, Zuo H R, Zhu S Y, *et al.* (2024). Fatigue degradation of wind turbines considering dynamic wake meandering effects [J]. *Engineering Structures*, 300.

Fluid Mechanics and Industrial Applications – Fang (Ed.)
© 2026 The Author, ISBN: 978-1-041-15371-9

PIV experimental study on the effectiveness of SDBD plasma aerodynamic excitation parameters on body force

Ningyuan Liu
Aerospace Technology Institute, China Aerodynamics Research and Development Center, China

Jinlong Li
Space Engineering University, Beijing, China

Yajun Liang, Guicai Fang and Hongliang Ji
Aerospace Technology Institute, China Aerodynamics Research and Development Center, Mianyang, China

ABSTRACT: When surface dielectric barrier discharge (SDBD) plasma generates aerodynamic excitation, the airflow in the flow field ionizes to form plasma, which generates body force under the action of electromagnetic fields. This is the main factor in forming plasma-induced jets and can directly affect the mixing performance of the jet. (PIV)is paper, particle image velocimetry(PIV)technology is used to measure the SDBD plasma-induced flow field under low pressureand the influence of plasma excitation parameters on plasma body force is studied. The results indicate that as the pulse frequency increases, the horizontal distribution range of the body force remains consistent, the primary action point shifts closer to the actuator, and the overall body force on the control body stays largely unchanged. As the duty cycle increases, the horizontal range of body force density first increases and then decreases, reaching a maximum at 50%, the main action point moves away from the actuator, and the resultant force on the control body increases. When the excitation voltage increases, the acting region of the body force remains unchanged, and both the maximum induced velocity and the resultant force on the control body increase linearly.

1 INSTRUCTIONS

The injector is an important component in engines for injecting propellant into the combustion chamber, which can directly affect the mixing performance of the injection jet, thereby influencing engine ignition and combustion efficiency. Most rocket engines used in spacecraft are low-thrust designs that feature a limited number of injector units in the combustion head, exhibit suboptimal propellant mixing capabilities, and possess a small combustion chamber volume. This leads to a brief residence and combustion time for the propellant, which hampers efficient combustion. The characteristics of surface dielectric barrier discharge (SDBD)Active flow control technology, which is a new kind of flow control method, include a quick reaction and a basic structure (Nie Wansheng, 2012). It utilizes plasma aerodynamic effects to create unsteady excitation in the airflow, resulting in substantial improvements in jet mixing performance. Additionally, when combined with the chemical kinetic effects of active particles in plasma, it enhances the ignition and combustion efficiency of low-thrust rocket engines. (Ju Y G, 2016)

In the last few years, the application of plasma active control technology at home and abroad has been widely popular (Wang J J, 2003). It has great application potential in

DOI: 10.1201/9781003679561-8

suppressing flow separation, controlling jet flow, engine efficiency enhancement, and flight control (Wu Yun, 2015). The basic structure of SDBD plasma actuators was proposed by the Roth team (1998). Enloe *et al.* (2004) displayed the characteristics of SDBD plasma induction through smoke line methods, pointing out that the induced flow is a near-wall jet flow, and measured the body force generated by the actuator using a balance, indicating that both excitation parameters and electrode structure have important effects on body force. With the development to PIV technology, Kotsonis *et al.* (2010) proposed using PIV to measure the flow field and obtaining body force by solving the N-S equations. In terms of excitation parameters, Benard *et al.* (2013) used LDV technology to measure sinusoidal wave excitation flow fields, studying the variation of body force within the sinusoidal wave cycle, and found significant differences in body force between the rising and falling half-cycles of sinusoidal discharge, with the latter having greater body force. Li *et al.* (2013) used PIV to measure the influence of excitation parameters on the flow field, pointing out that when voltage increases, the maximum SDBD-induced velocity increases linearly, and when duty cycle increases, velocity also increases, while pulse frequency changes produce nonlinear effects. Li Yiwen (2009) also confirmed using a balance that when excitation voltage increases, body force shows an obvious linear relationship, while excitation frequency has little effect on body force. Zhou Penghui (2013) used PIV to measure the body force generated by SDBD plasma, pointing out that changes in excitation voltage affect both the magnitude and distribution of body force.

When SDBD plasma generates aerodynamic excitation, the airflow in the flow field ionizes to form plasma, which generates body force under electromagnetic field action. Body force is the main factor in forming plasma-induced jets and can directly affect flow control performance. Therefore, analyzing the variation of body force with excitation parameters and understanding the influence of each parameter on actuator performance is the basis for conducting flow control research. This paper uses PIV technology to measure the flow field caused by SDBD plasma under various excitation parameter circumstances, and then studies the impact of plasma excitation parameters on the force of the plasma body.

1.1 *Experimental system*

In low pressure, the SDBD plasma-induced flow field is measured by PIV technology. Figure 1 illustrates the experimental system, which primarily consists of the test chamber along with its control system, the PIV system, the excitation power supply, and the SDBD actuator. The test chamber is cylindrical with an internal diameter of 600 mm and a length of 800 mm; auxiliary equipment includes gas filling valves, fine-tuning valves, temperature sensors, nitrogen cylinders, etc.; the test chamber pressure is controlled by a vacuum chamber control cabinet, the vacuum pump has a pumping speed of 14 L/s, and the pressure is adjusted according to experimental conditions with an accuracy of 100 Pa. Two high voltage terminals are introduced into the test chamber by the excitation power supply; the terminals are copper columns with polytetrafluoroethylene insulation layers on the outside of them. In order to control the phenomena of creeping discharge, the surface of the insulation layer is treated with threads. With the help of O-rings, the copper column and insulation layer are firmly bonded and sealed. The test chamber is equipped with 200 mm diameter quartz observation windows on both the top and side; the top window facilitates PIV laser transmission, while the side window is designed for CCD camera photography.

An Nd: YAG dual-pulse laser, Synchronization controller, CCD camera, computer, image processing software, and CCD camera are all included in the PIV system. The laser pulse repetition frequency during the experiment is 5 Hz. The CCD camera is a Kodak scientific-grade chip camera with a resolution of 2048×2048 px/in. Smoke produced by burning smoke cakes serves as tracer particles in the experiment. After the particles stabilize, the actuator begins to work, and PIV image capture is initiated. Image post-processing uses MicroVec V3.1.1 calculation software with a resolution of 2.62 mm.

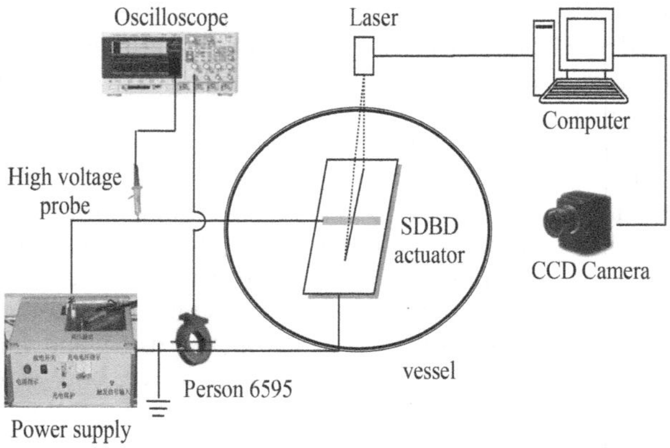

Figure 1. Experimental system.

Cu foil is used to fabricate the actuator electrodes, and the exposed electrode is used, the electrode is embedded, and the electrode spacing is changed based on the experimental requirements. The dielectric layer is a 1.5 mm thick epoxy resin board with a black matte paint coating on the surface to avoid laser reflection from the dielectric layer surface affecting PIV measurements. The actuator is placed horizontally in the test chamber, with the exposed electrode on the left side of the observation window where the camera is located, and the embedded electrode on the right side.

The excitation power supply uses the HFHV30-1 type high-voltage power supply created by the Institute of Electrical Engineering, Chinese Academy of Sciences, with an output voltage of 0-30 kV and a sinusoidal waveform, having both continuous and pulse working modes. In continuous operating mode, the output frequency is 1-50 Hz, and in pulse mode, both pulse frequency and duty cycle are adjusted with minimum steps of 10 Hz and 10% respectively.Voltage is measured using an Agilent N2771B high-voltage probe, current is measured with a Pearson current coil 6595, and voltage and current measurement results are shown and recorded using an Agilent DSO3024 oscilloscope.

This study uses pulse working mode, which is an unsteady excitation mode. In continuous sinusoidal wave excitation, a certain discharge period and duty cycle can form unsteady excitation.In pulse mode, the continuous operating frequency is the carrier frequency w; the pulse frequency f_p is associated with a single complete discharge period T_p, meaning $T_p =\frac{1}{f_p}$;and the duty cycle D is defined as the ratio of the effective discharge time T_1 to T_p, expressed as $D =\frac{T_1}{T_p}$.The waveform schematic diagram under pulse discharge mode is shown in the Figure 2.

Figure 2. Low-pressure test chamber.

1.2 *Plasma body force solution method*

The SDBD-induced jet measured by PIV in the experiment is a two-dimensional flow field. Because the jet flow velocity induced by SDBD is low, the flow can be regarded as incompressible. Therefore, the body force distribution can be solved according to the two-dimensional incompressible N-S equation:

$$\frac{\partial \boldsymbol{U}}{\partial t} + \boldsymbol{U} \cdot \nabla \boldsymbol{U} - \nu \nabla^2 \boldsymbol{U} = -\nabla p + \frac{\boldsymbol{F}}{\rho} \tag{1}$$

Where U is the flow field velocity vector, v is the gas dynamic viscosity, ρ is the gas density, F is the body force, and p is the gas pressure.

The flow field measured by PIV is the average flow field after the actuator stabilizes, so the flow can be considered steady, and the pressure effect can be ignored. Equation (1) becomes:

$$\boldsymbol{U} \cdot \nabla \boldsymbol{U} - \nu \nabla^2 \boldsymbol{U} = \frac{\boldsymbol{F}}{\rho} \tag{2}$$

Decomposing Equation (2) in the x and y directions respectively, the solution formulas for body force density in the x and y directions can be obtained:

$$f_x = \rho\left(u\frac{\partial u}{\partial x} + v\frac{\partial u}{\partial y}\right) - \mu\left(\frac{\partial^2 u}{\partial x^2} + \frac{\partial^2 u}{\partial y^2}\right) \tag{3}$$

$$f_y = \rho\left(u\frac{\partial v}{\partial x} + v\frac{\partial v}{\partial y}\right) - \mu\left(\frac{\partial^2 v}{\partial x^2} + \frac{\partial^2 v}{\partial y^2}\right) \tag{4}$$

$$f = \left(f_x^2 + f_y^2\right)^{\frac{1}{2}} \tag{5}$$

In this equation, u and v represent the decomposed velocities in the x and y directions, respectively, while μ denotes the dynamic viscosity of the gas. Fors DBD, the resulting body power on the control body is:

$$F_x = \oiint_S f_x dxdy \tag{6}$$

$$F_y = \oiint_S f_y dxdy \tag{7}$$

$$F = \left(F_x^2 + F_y^2\right)^{\frac{1}{2}} \tag{8}$$

2 INFLUENCE OF PLASMA EXCITATION PARAMETERS

Pulse frequency, duty cycle, excitation voltage, and other power supply main excitation characteristics are listed. The SDBD plasma body force changes with excitation parameters, causing corresponding changes in the induced flow field. When engines ignite in the complex high-altitude environment, the ambient pressure is usually lower than atmospheric pressure. Most current research on SDBD plasma aerodynamic excitation characteristics focuses on atmospheric pressure conditions, so it is necessary to study the aerodynamic excitation characteristics of SDBD plasma under low pressure. During the experiment, the exposed electrode and embedded electrode widths of the SDBD actuator are kept at 9 mm each, the electrode gap is 0, and the overlapping length of the two electrodes is 10 cm. The power supply carrier frequency w = 10kHz is maintained during the experiment. When changing the pulse frequency, Vp-p = 8kV and D = 50%are kept; when changing the duty cycle, Vp-p = 8kV and fp = 50Hz are kept; when changing the excitation voltage, fp and D are 50Hz and 50%respectively. At 70.1 kPa, the test chamber pressure is kept, and at 12°C, the temperature is maintained.

2.1 *Pulse frequency*

The induced flow field is depicted in Figure 3 when the actuator runs stably. The connection point of the two electrodes is located at x = 0mm.Under normal pressure, the SDBD-induced jet is basically in line with the jet morphology of the jet. That is, the plasma attracts the gas above the embedded electrode to the wall surface and then pushes it out, thus forming a wall jet. Under low pressure, discharge is more likely to occur on both sides of the exposed electrode, so after the gas is attracted to the wall surface, a small part forms a weak negative x-direction wall jet. Under the current excitation parameters, as the pulse frequency increases, the flow field morphology changes from a single vortex to two vortices; as the pulse frequency continues to increase, the right vortex of the flow field gradually moves to the right until it dissipates, and the measured maximum induced velocity slightly increases.

Figure 4 shows the distribution of volume force density f_x and f_y when the duty cycle is 50% and f_p = 90Hz. The white solid line box serves as the control body for solving the resultant volume force. The resultant force control body in this chapter is the area within this white solid line box. Under these power parameters, the positive values of f_x are relatively concentrated near the wall surface at x = 20mm, while negative values of f_x appear at x = −5mm and 40mm.Because fx has a negative value at x = 40 mm because of the friction between the wall surface and airflow, ignoring the pressure gradient that results from the low flow velocity close to the wall surface is also what causes it to be negative. The negative value appearing at x =

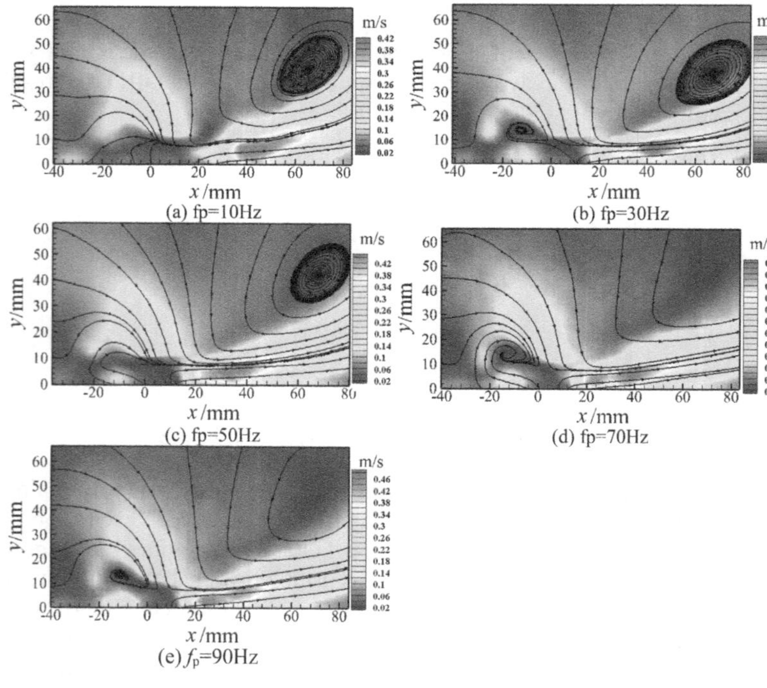

Figure 3. Induced flow field under different pulse frequencies.

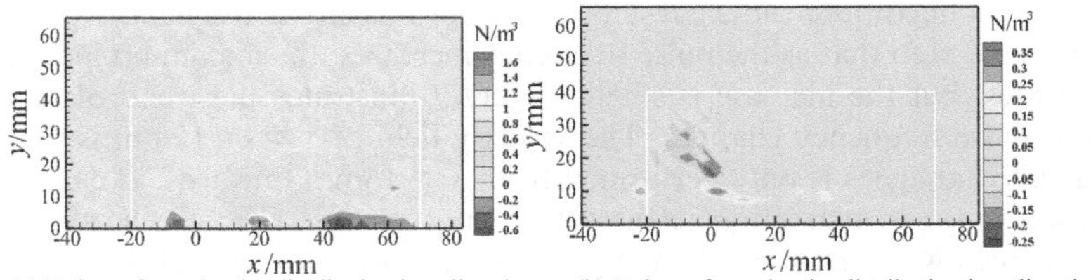

(a) Volume force density distribution in x direction (b) Volume force density distribution in y direction

Figure 4. Volume force component density distribution at 8kV-90Hz-50%.

-5mm is caused by the discharge on both sides of the exposed electrode generating a leftward volume force, thus causing a weak leftward jet. Compared to f_x, the value of f_y is relatively small, mainly distributed above the actuator, and slightly away from the wall surface. The force in the y direction mainly attracts the airflow above the SDBD to the near-wall surface, and then pushes it to both sides through the x direction force to form a jet.

The distribution of f under various pulse frequencies is depicted in Figure 5. where the 0.2N/m3 contour line is represented by the black solid line. It can be seen that when f_p = 10Hz, the maximum value of f appears near the wall surface at x = 35mm, the horizontal distribution range is about 40mm, and the acceleration effect on the airflow at the maximum value position is relatively obvious. As the pulse frequency rises, the distribution area of the volume force density f shifts closer to the actuator. Although the horizontal influence range remains largely constant, the position of the maximum value shifts toward the actuator, and the maximum value itself continues to increase. Where fp = 90Hz, at x = 20mm, the maximum value of f appears close to the wall surface. Compared to f_p = 10Hz, the maximum value position of f moves about 15mm closer to the actuator direction, and the horizontal distribution range basically remains at 40mm.

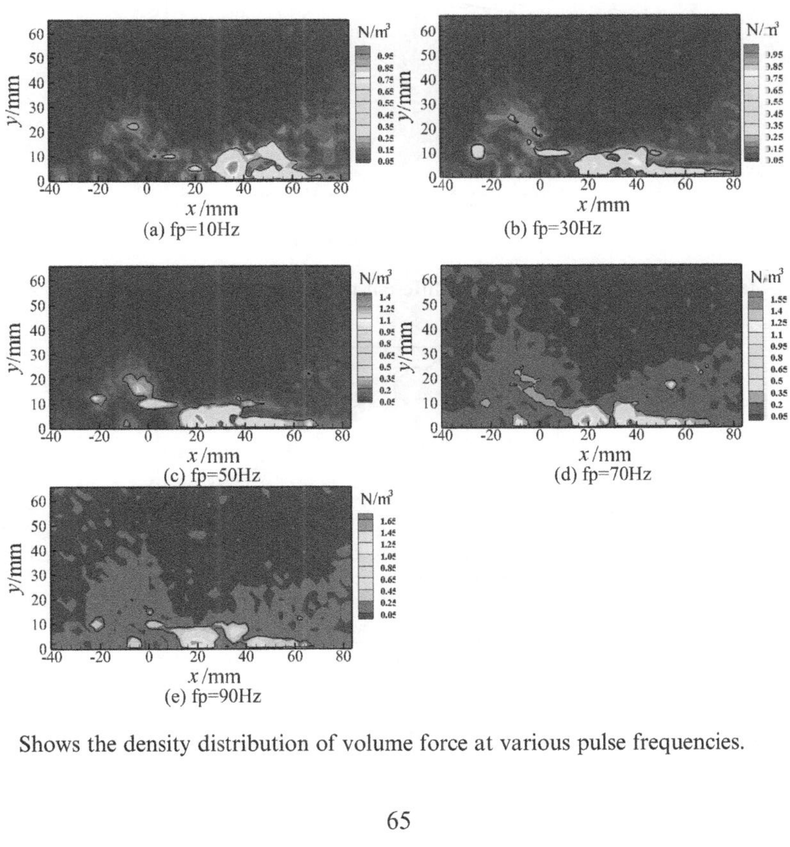

(a) fp=10Hz (b) fp=30Hz (c) fp=50Hz (d) fp=70Hz (e) fp=90Hz

Figure 5. Shows the density distribution of volume force at various pulse frequencies.

The change in maximum induced velocity at various pulse frequencies is depicted in Figure 6. It can be seen that as the pulse frequency increases, the maximum induced velocity slightly increases, but the increase is small. Figure 7 illustrates the trend of maximum jet velocity with pulse frequency changes. The velocity field at x = 0–15mm is not fully captured, so detailed analysis is only performed from x = 15mm onwards. It can be seen that although the change in maximum jet velocity is not obvious, the actuator's acceleration process of the airflow is different: when the pulse frequency is low, the airflow acceleration is relatively lagging, and the main jet position is relatively far from the actuator; as the pulse frequency increases, the airflow acceleration zone moves closer to the actuator, and the main jet moves toward the actuator direction. When f_p = 70Hz and 90Hz, the airflow at x = 25mm has already completed acceleration.

Regarding the range of volume force action, Figure 8 shows the distribution of volume force density f in the airflow layers at y = 1mm and 5mm. The peak occurs upstream of the streamline direction at the location of maximum induced velocity, which is consistent with the conclusion in literature (Zhou Penghui, 2013). Volume force is the power source of gas jet and determines the acceleration state of the airflow. As the pulse frequency increases, the volume force peak on the right side of the actuator continues to increase and moves closer to the actuator. For the airflow layers at y = 1mm and 5mm, the maximum volume force density f_{max}, when the pulse frequency is 10Hz, f_{max} appears at x = 35mm in the y = 5mm airflow layer, reaching 1.0 N/m^3; when the pulse frequency is 90Hz, f max appears at x = 22mm in the y = 1mm airflow layer, reaching 1.4N/m^3. This indicates that when the pulse frequency is low, the volume force action range is relatively high, while as the frequency increases, the main action point of the volume force is closer to the wall surface.

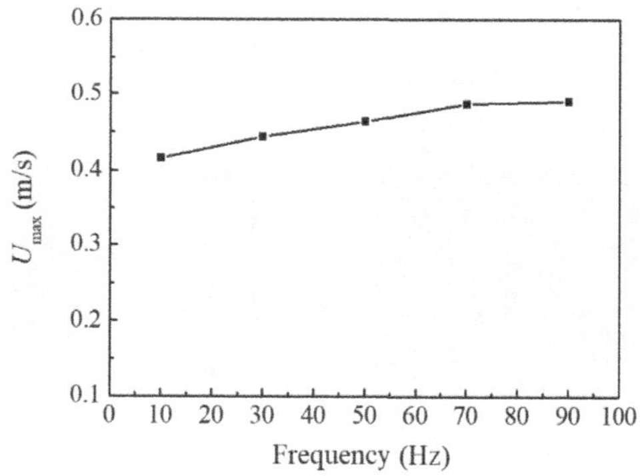

Figure 6. Maximum velocity change under different pulse frequencies.

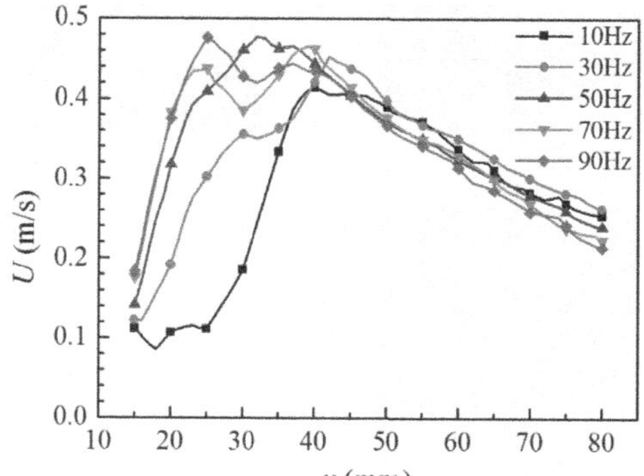

Figure 7. Maximum velocity distribution at different x positions under different pulse frequencies.

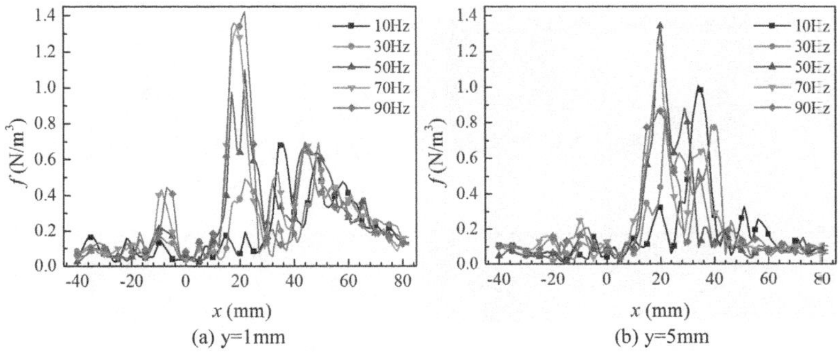

Figure 8. Shows the density distribution of volume force at various heights with varying pulse frequencies.

Figure 9 shows the trend of resultant volume force F_x and F y on the control body with pulse frequency changes. It can be seen that F_x is much greater than F_y, indicating that the horizontal momentum exchange between plasma and airflow dominates. As the pulse frequency increases, F_x on the control body first rises and then decreases, but the numerical change is not obvious. This also shows that the resultant volume force F on the control body does not change significantly with increasing pulse frequency. Overall, increasing the pulse frequency mainly causes the volume force action point to move closer to the actuator and wall surface, while the resultant force does not change significantly, resulting in no obvious change in maximum jet velocity, but the vortex and airflow acceleration processes in the flow field do change.

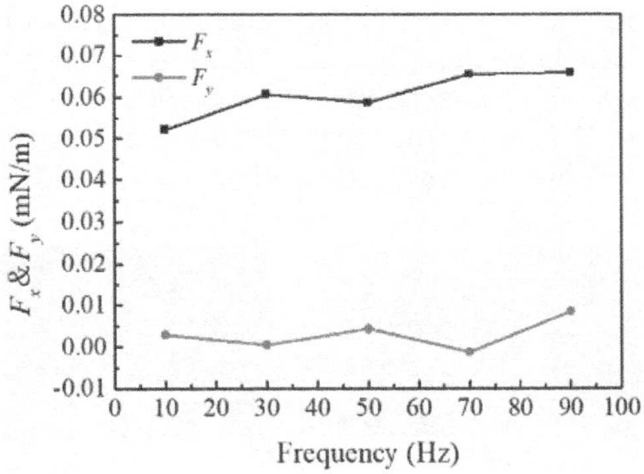

Figure 9. Resultant volume force under different pulse frequencies.

2.2 *Duty cycle*

Figure 10 shows when the actuator stabilizes, the induced flow field after the duty cycle is changed. Discharge continues to happen on both sides of the exposed electrode, resulting in the formation of two induced vortices in the flow field. When D = 10%, the maximum induced velocity is about 0.18m/s, and the center of the right vortex is approximately located above x = 40mm. As the duty cycle increases, the maximum induced velocity also increases, and the right vortex slightly moves away from the actuator to the right. When D = 90%, the maximum induced velocity is about 0.42m/s, the vortex center is approximately above x = 75mm, and it has a tendency to dissipate.

Figure 11 illustrates the distribution of f across various duty cycles, with the solid black line representing the 0.2 N/m^3 contour line. When D = 10%, the distribution range of f is relatively small and close to the actuator, with a horizontal distribution range of about

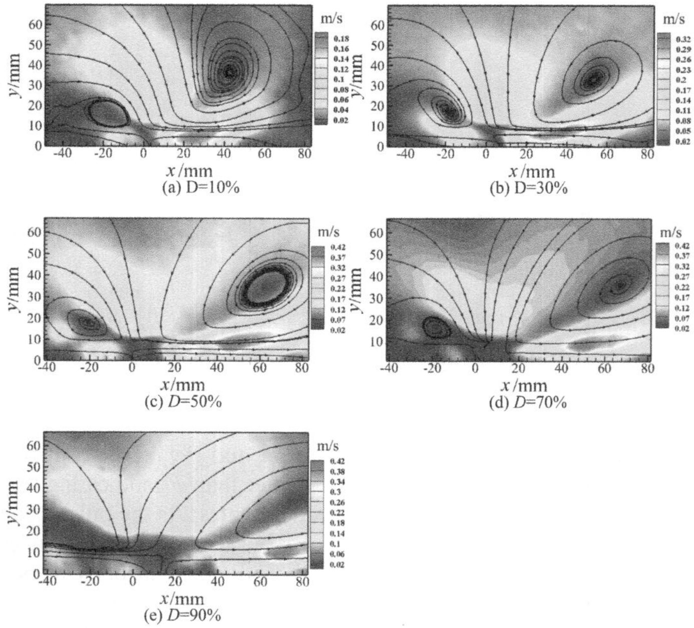

Figure 10. Induced Flow field under different duty cycles.

10mm. When D changes between 10%–50%, the main distribution area off gradually moves away from the actuator, and the distribution range continues to increase, reaching about 40mm at D = 50%; when D changes between 50%–90%, the main distribution area off continues to move away from the actuator, but the horizontal influence range decreases, reaching about 15mm at D = 90%. The maximum value of f increases with the duty cycle, about 0.22 N/m^3 at D = 10%, and reaching 1.05 N/m^3 at D = 90%.

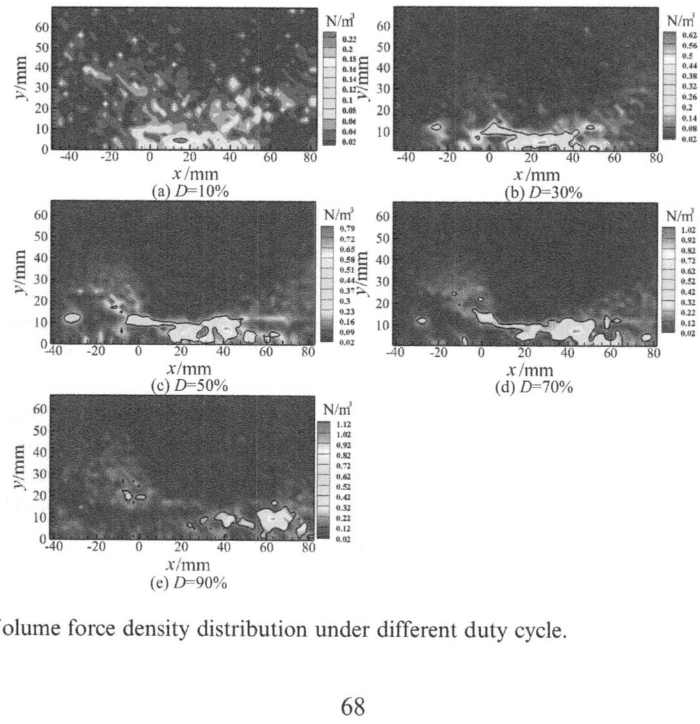

Figure 11. Volume force density distribution under different duty cycle.

The change in maximum induced velocity with duty cycle is depicted in Figure 12. As the duty cycle D varies from 10% to 30%, the maximum velocity rises in discrete steps; however, when D shifts from 30% to 90%, the maximum velocity tends to increase almost linearly. Figure 13 shows the change in volume force F_x and F_y on the control body with duty cycle. When the duty cycle low, both Fxand Fyare small, because the injected electrical energy is low at this time, the generated power is also small, and it is concentrated near the actuator. As the duty cycle increases, F_x increases approximately linearly between D = 30%–90%, which is closely related to the enhanced energy injection with increasing duty cycle. The increase in F_x makes the resultant volume force F increase, thereby increasing the maximum velocity.

Regarding the phenomenon that the growth trend of volume force resultant between D = 10%–30% is smaller compared to the growth trend of maximum velocity, Figure 14 shows the f values in the airflow layers at y = 1mm, 5mm, and 9mm. The volume force generated by discharge on the left side of the actuator is weak and increases with the duty cycle, but the acting height is still limited. The volume force density distribution in the airflow layers at y = 1mm and 5mm has similar trends, but the f values at y = 5mm are all larger than those at y = 1mm. The upper airflow layer receives greater volume force than the lower layer, making the upper airflow faster than the lower airflow, which has a suction effect on the lower gas. When the duty cycle is below 50%, the maximum value of volume force density fmax appears at y = 5mm.

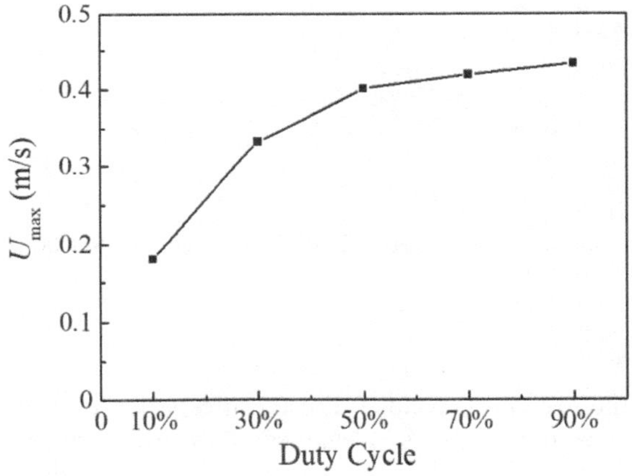

Figure 12. Maximum velocity change under different duty cycle.

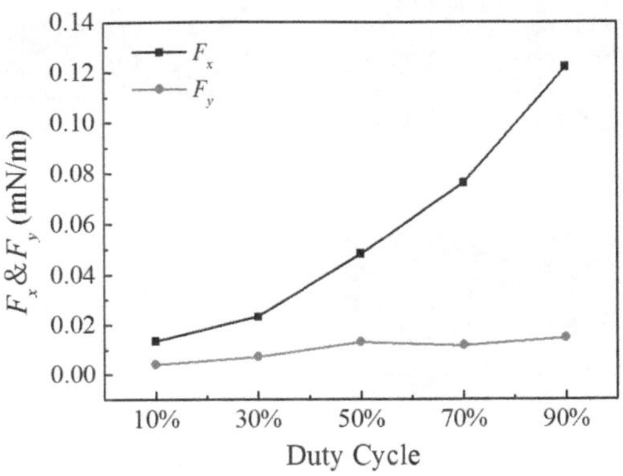

Figure 13. The volume force that was produced under various duty cycles.

As the duty cycle continues to increase, it can be seen from Figure 14(c) that f max appears in the airflow layer at y = 9mm. When D = 10%, the volume force density at y = 9mm is already close to 0, but when D = 30%, there is still a certain volume force density distribution at y = 9mm. Therefore, when D = 10%, the gas acceleration process is greatly affected by the wall surface, while when D = 30%, the acting height increases, thus significantly increasing the maximum velocity with little increase in volume force. All things considered, the acting height of the volume force rises with the duty cycle as well. The increase in duty cycle makes the f distribution area move away from the actuator, while the resultant force increases, acting on gas farther away, making the influence range of volume force more extensive.

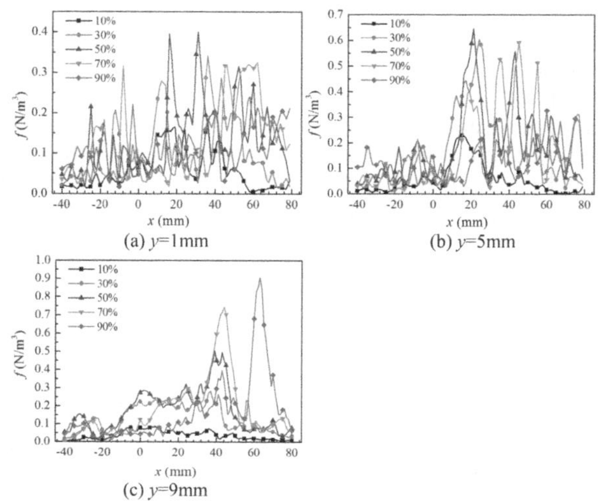

Figure 14. Volume force density distribution in different height airflow layers under different duty cycles.

2.3 *Excitation voltage*

The magnitude of the excitation voltage directly affects the level of input energy, mainly reflected in the changes in the numerical values of various variables. When changing the excitation voltage, Figure 15 shows the change in volume force density distribution f with

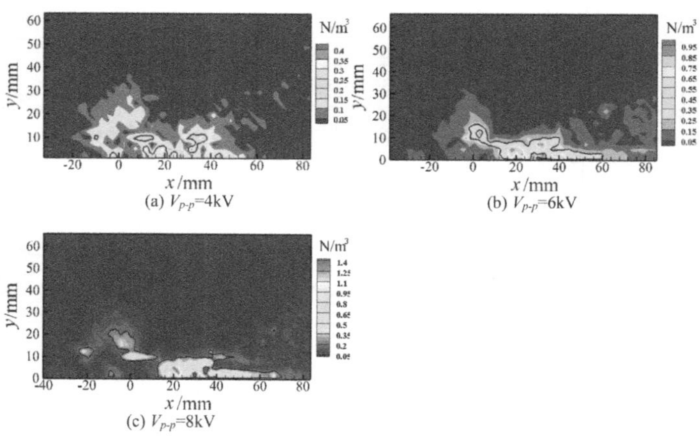

Figure 15. Shows the density distribution of volume force at various excitation voltages.

voltage, where the black solid line represents the $0.2N/m^3$ contour line. The main concentration area of volume force density has almost no change, all relatively concentrated near x = 40mm, and the values increase with voltage.

To study the specific influence of excitation voltage on the numerical values of various variables, Figure 16 shows the change in maximum induced velocity under different voltages. It can be seen that the maximum induced velocity has a strong linear relationship with the excitation voltage, which is similar to the conclusion in literature (Li F, 2013). Figures 17 and 18 illustrate the correlation between volume force and excitation voltage. It can be seen that excitation voltage directly affects the actuator power, and as the actuator power increases linearly, the volume force and voltage also show a strong linear relationship.

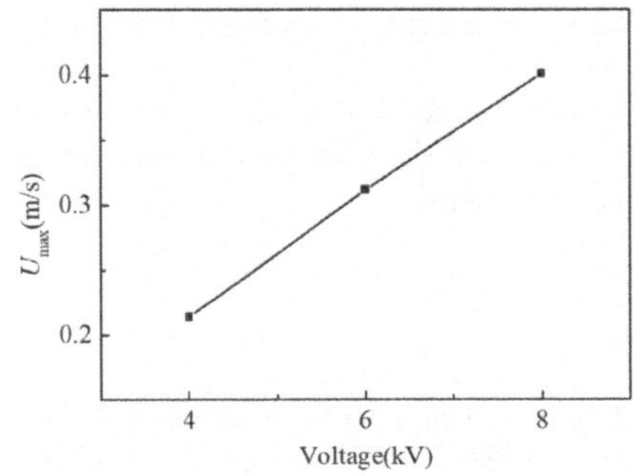

Figure 16. Shows the maximum velocity that was produced at various excitation voltages.

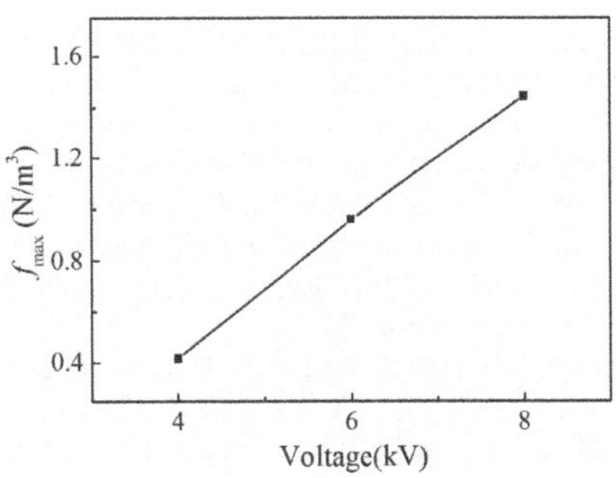

Figure 17. Displays the maximum volume force density corresponding to various excitation voltages.

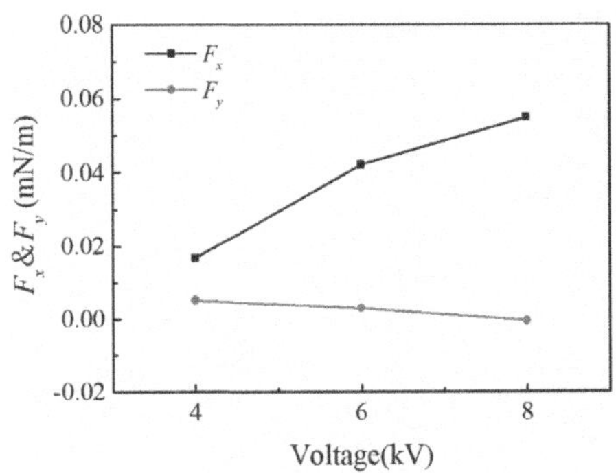

Figure 18. Illustrates the resultant volume force for various excitation voltages.

3 CONCLUSION

This chapter employs PIV experimental technology to assess the flow field induced by SDBD plasma at 7 kPa, calculates the volume force produced by the plasma, and examines how excitation and structural parameters affect SDBD plasma aerodynamic excitation. The main conclusions are as follows:

(1) The main point of the motion toward the actuator direction and the resulting volume force on the control body basically stays unchanged when the pulse frequency increases because the horizontal distribution range of the volume force stays the same.
(2) When the duty cycle increases, the horizontal range of volume force density first increases and then decreases, reaching a maximum at 50%, the main action point gradually moves away from the actuator, and the resultant force on the control body increases.
(3) When the excitation voltage increases, the main acting region of the volume force does not change significantly, and both the maximum induced velocity and the resultant force on the control body increase linearly.

REFERENCES

Benard N, Debien A, Moreau E. (2013). Time-dependent volume force produced by a non-thermal plasma actuator from experimental velocity field[J]. *Journal of Physics D Applied Physics*, 46(24):245201.

Enloe C L, Mclaughlin T E, Vandyken R D, *et al.* (2004). Mechanisms and Responses of a Single Dielectric Barrier Plasma[J]. *AIAA Journal*, 42(3):589–594.

Ju Y G, Lefkowitz J K, Reuter C B, *et al.* (2016). Plasma assisted low temperature combustion[J]. *Plasma Chemistry and Plasma Processing*, (36):85–105.

Kotsonis M, Veldhuis L, Ghaemi S, *et al.* (2010). Experimental study on the body force field of dielectric barrier discharge actuators[R]. AIAA Paper 2010–4630.

Li F, Gao C, Zheng B, *et al.* (2013). Aerodynamic flow acceleration with one atmosphere uniform induced by single and multiple DBD plasma actuation based on PIV[R]. AIAA Paper, 2013–2746.

Li Yiwen, Li Yinghong, Zhou Zhangwen, *et al.* (2009). Experimental study on volume force of plasma aerodynamic excitation[J]. *Journal of Experimental Fluid Mechanics*, 23(2):82–86.

Nie Wansheng, Cheng Yufeng, Che Xueke. (2012). Research progress in dielectric barrier discharge plasma flow control[J]. *Advances in Mechanics*, 42(6):722–734.

Roth J R. Sherman D M, Wilkinson S P. (1998). Boundary layer flow control with one atmosphere uniform glow discharge surface surface plasma[R]. AIAA Paper 1998–0328.

Wang J J, Choi K S, Feng L H, *et al.* (2003). Bifurcating and blooming jets[J]. *Annual Review of Fluid Mechanics*, 35(35):295–315.

Wu Yun, Li Yinghong. (2015). Research progress and prospects in plasma flow control[J]. *Acta Aeronautica et Astronautica Sinica*, 2015, 36(2):381–405.

Zhou Penghui. (2013). Simulation and experimental research on plasma flow control in near space[D]. *Beijing: Academy of Equipment*, 2013.

Fluid Mechanics and Industrial Applications – Fang (Ed.)
© 2026 The Author, ISBN: 978-1-041-15371-9

Experimental study on constant rheology water-based drilling fluids with different types of base slurries

Zhaochuan Li, Can Liu, Xiaogang Wang and Guangyi Wan
Sinopec Shengli Oilfield Branch Technical Testing Center, Shandong Province, China

ABSTRACT: With increasing seawater depth, the temperature progressively decreases, reaching approximately 4°C at significant depths at the seabed mud line. In such low-temperature conditions, the viscosity and shear force of the drilling fluid experience a considerable increase. Conversely, during drilling operations with circulation, the temperature of the drilling fluid in the riser pipe rises, leading to a reduction in viscosity. The key component in marine deepwater water-based drilling fluids is the base slurries, which is primarily responsible for enhancing viscosity and shear, reducing fluid loss, and ensuring wellbore integrity and wall construction. The preparation of the drilling fluid involves initially mixing the base slurries with water in a predetermined ratio to create a base slurry, to which various treatment agents are subsequently added. It is essential to maintain the stability of the base slurry's rheological properties within a temperature range of 4°C to 65°C to ensure the constant rheology of marine deepwater water-based drilling fluids.

1 CLAY ANALYSIS FOR SLURRY PREPARATION

At present, the clays predominantly utilized for the formulation of drilling fluids are categorized into four types: sodium bentonite, calcium bentonite, sepiolite, and attapulgite. (Arkadiusz Jo zefczak, 2011; Hanyi Zhong, 2011; Rafael Roa, 2012; Robert Schlemmer, 2010; Shuo Wu, 2024; Young Gun Ko, 2009) Bentonite, the earliest adopted clay for this purpose, is favored due to its high montmorillonite content, efficient slurry generation, and extensive availability, making it a widely employed choice in drilling operations globally. (Basim Abu-Jdayil., 2011; Hege E, 2001) Sepiolite and attapulgite are noted for their high-temperature tolerance and salt resistance, and when combined with fluid loss additives, they are suitable for applications in saturated brine and brine-based drilling fluids. Moreover, there is ongoing research into the potential use of kaolinite as a slurry preparation material. (Xin Zhang, 2023)

To examine the potential relationship between the comprehensive mineral composition of the clays used for slurry preparation and the rheological properties of the resulting base slurry within the temperature range of 4°C to 65°C an X-ray diffraction analysis was performed to ascertain the complete mineral composition of various clay samples. The findings of this analysis are detailed in Table 1.

This study aims to determine whether there is a correlation between the clay mineral composition of the slurry soil and the rheological properties of the base slurry within a temperature range of 4°C to 65°C. X-ray diffraction analysis was performed on the complete mineral composition of the clay in each slurry soil sample, and the results are presented in Table 2.

DOI: 10.1201/9781003679561-9

Table 1. Mineral compositions of bentonites.

Type	Quartz	Orthoclase	Plagioclase	Calcite	Dolomite	Total clay minerals
Calcium bentonite①	30	4	–	7	–	59
Calcium bentonite②	26	6	14	11	–	43
Sodium bentonite①	29	–	–	15	–	56
Sodium bentonite②	36	7	–	–	–	56
Sepiolite①	9	8	17	15	23	25
Sepiolite②	21	–	–	–	–	79
Aattapulgite①	11	5	–	–	23	61
Aattapulgite②	24	4	4	–	3	64
Kaolinite①	3	–	12	23	4	58
Kaolinite②	43	13	25	4	3	12

Table 2. Clay compositions of bentonites.

Type	(K)	(Ch)	(I)	(S)	(P)	(I/S)	(C/S)
Calcium bentonite①	3	2	2	93	0	0	0
Calcium bentonite②	5	3	5	87	0	0	0
Sodium bentonite①	3	1	1	0	0	96	0
Sodium bentonite②	2	1	7	90	0	0	0
Sepiolite①	0	0	3	0	97	0	0
Sepiolite②	0	0	0	0	84	16	0
Aattapulgite①	0	0	6	0	51	43	0
Aattapulgite②	0	0	0	0	100	0	0
Kaolinite①	11	8	8	0	0	73	0
Kaolinite②	11	1	46	0	0	42	0

This study aims to determine whether there is a correlation between the oxide composition of the slurry soil and the rheological properties of the base slurry within a temperature range of 4°C to 65°C. X-ray fluorescence analysis was performed on the oxide composition of each slurry soil sample, and the results are presented in Table 3.

Table 3. Chemical compositions of bentonites.

Type	SiO_2	Al_2O_3	Fe_2O_3	CaO	MgO	TiO_2	K_2O	Na_2O	SO_3
Calcium bentonite①	58.02	18.02	8.66	3.21	1.82	1.24	0.80	0.23	0.03
Calcium bentonite②	56.18	17.68	5.57	6.33	1.54	0.58	1.39	0.94	0.04
Sodium bentonite①	53.34	16.34	8.81	4.74	1.92	1.25	0.74	3.60	0.02
Sodium bentonite②	53.34	16.34	8.81	4.74	1.92	1.25	0.74	3.60	0.02
Sepiolite①	46.91	3.69	1.26	11.07	17.11	0.20	0.97	0.50	0.04
Sepiolite②	59.92	10.25	5.33	1.26	10.89	0.56	0.97	0.11	0.02
Aattapulgite①	52.26	10.69	5.82	6.80	9.05	0.81	1.05	0.13	-
Aattapulgite②	62.32	12.28	6.70	2.00	5.93	0.99	1.68	0.16	0.12
Kaolinite①	50.82	24.25	0.83	10.31	2.54	0.07	0.80	1.07	0.02
Kaolinite②	69.97	15.33	0.63	1.85	0.88	0.05	3.30	2.80	0.01

2 EXPERIMENTAL PROCEDURES

(1) A base slurry with a clay concentration of 4.0% by weight was prepared, and its pH was standardized to 9 using either 5M sodium hydroxide or 5M hydrochloric acid.
(2) The base slurry was then subjected to a static condition at 4°C for 6 hours, followed by rheological testing to determine its apparent viscosity, plastic viscosity, initial and final shear, and API filter loss at 4°C. Subsequently, the slurry was heated using a rheometer's heating jacket, and the same rheological properties were measured at 25°C and 65°C.
(3) The base slurry was transferred into an aging vessel and subjected to a high-temperature rolling process at 150°Cfor 16 hours. Following the aging process, the slurry was allowed to cool to ambient temperature.
(4) The aged base slurry was then placed back at 4°C for 6 hours, and its apparent viscosity, plastic viscosity, initial and final shear, and API filter loss were again tested at 4°C. The slurry was reheated to 25°C and 65°C using the rheometer's heating jacket, and the rheological properties were measured at these temperatures.

3 EXPERIMENTAL RESULTS ANALYSIS

(1) The experimental findings, as presented in Table 4, indicate that the calcium bentonite-based slurries exhibit stable performance across a broad temperature range of 4°C to 65°C. Notably, these slurries maintain relatively low viscosities, with minimal changes in their properties following the aging process. The slight increase in apparent and plastic viscosities at lower temperatures suggests a robust stability profile. The primary constituents of the calcium bentonite slurries are clay minerals, which include trace amounts of quartz, potassium feldspar, and calcite. The predominant clay mineral is montmorillonite, with lesser quantities of illite, chlorite, and kaolinite. Montmorillonite's unique 2:1 layer structure, comprising two silicate tetrahedral layers and one octahedral layer, allows for interlayer expansion due to the absence of K^+ ions, enabling free water entry and lattice swelling. This, coupled with its weak interlayer bonding, contributes to its excellent dispersibility and, consequently, the rheological properties of the slurry. In contrast, illite, also a 2:1 layer mineral, has a more rigid lattice structure due to K^+ ion bonding, resulting in lower dispersibility and expansibility compared to montmorillonite. Hence, from a mineralogical standpoint, montmorillonite plays a pivotal role in determining the dispersibility of bentonite, which is a key factor influencing the rheological behavior of the slurry.

Table 4. Properties of Ca-bentonites.

Number	Condition	T/°C	AV/mPa·s	PV/mPa·s	YP/Pa	Gel/Pa/Pa	FL_{API}/mL
①	Before aging	65	1.50	1.50	0	0/0	—
		25	1.50	1.50	0	0/0	
		4	1.75	1.50	0.25	0/0	
	After aging	65	1.25	0.50	0.75	0/0	—
		25	1.25	0.50	0.75	0/0	
		4	1.50	1.50	0	0/0	
②	Before aging	65	1.50	1.50	0	0/0	—
		25	1.50	1.50	0	0/0	
		4	1.75	1.50	0.25	0/0	
	After aging	65	1.25	1.00	0.25	0/0	—
		25	1.25	1.00	0.25	0/0	
		4	1.50	1.50	0	0/0	

The crystal structure of montmorillonite consists of two layers of silicon-oxygen tetrahedra with an intervening layer of aluminum-oxygen (hydroxyl) octahedra. The oxygen atoms at the vertices of the tetrahedra are shared with the octahedra and face inward. This arrangement extends infinitely along the a-axis and b-axis, while along the c-axis, the layers are stacked with a defined interlayer spacing, forming a layered silicate structure. Lattice substitution results in the replacement of Si^{4+} in the tetrahedral layer by Al^{3+} and the replacement of Al^{3+} in the octahedral layer by divalent cations such as Ca^{2+}, Mg^{2+}, and Fe^{2+}, imparting a permanent negative charge to the lattice. When the negative charge of montmorillonite is neutralized by Ca^{2+}, calcium montmorillonite is formed. The high concentration of Ca^{2+} in calcium montmorillonite leads to a reduction in the thickness of the electric double layer between particles, causing the attractive forces to dominate over the repulsive forces. Consequently, the dispersibility of montmorillonite is significantly reduced, resulting in a lower viscosity for the calcium bentonite-based slurry.

(2) The performance of the two sodium bentonite-based slurries within the temperature range of 4°C to 65°C is shown in Table 5.

Table 5. Properties of Na-bentonites.

Number	Condition	T/°C	AV/mPa·s	PV/mPa·s	YP/Pa	Gel/Pa/Pa	FL_{API}/mL
①	Before aging	65	2.00	1.50	0.50	0/0.50	22.0
		25	2.75	2.00	0.75	0/0.50	20.0
		4	3.50	3.00	0.50	0.25/0.50	19.0
	After aging	65	3.00	2.50	0.50	0/0.25	40.0
		25	3.50	3.00	0.50	0/0.25	32.0
		4	4.00	3.50	0.50	0.50/1.50	22.0
②	Before aging	65	6.00	3.00	3.00	1.00/1.50	27.0
		25	7.75	4.00	3.75	1.50/2.25	24.0
		4	8.75	6.00	2.75	1.50/2.25	22.0
	After aging	65	3.75	3.00	0.75	0/0	35.0
		25	4.25	4.00	0.25	0/0	31.0
		4	6.50	5.00	1.50	0.50/0.75	29.0

The analysis of Table 5 reveals that the sodium bentonite-based slurries exhibit a higher viscosity with insignificant changes in properties pre- and post-aging. The decrease in temperature leads to a notable increase in both apparent and plastic viscosities. The primary constituents of the sodium bentonites remain clay minerals, supplemented by trace amounts of quartz, potassium feldspar, and calcite. Montmorillonite and illite/montmorillonite mixed layers are predominant in the clay mineral fraction, while the proportions of illite, chlorite, and kaolinite are relatively low. The mineral composition of sodium bentonite parallels that of calcium bentonite, differing primarily in the formation of sodium montmorillonite due to the lattice substitution of negative charges by Na^+ ions. In line with the electric double layer theory, the thickness of the double layer is inversely proportional to the ion valence. Consequently, the sodium montmorillonite, which adsorbs monovalent sodium ions, possesses a thicker double layer compared to the calcium montmorillonite, which adsorbs divalent calcium ions. This results in a higher ζ potential, stronger repulsive forces, and enhanced hydration and dispersion capabilities, leading to a higher viscosity and improved colloidal stability in the sodium bentonite-based slurry.

(3) The performance of the two sepiolite-based slurries within the temperature range of 4°C to 65°C is presented in Table 6.

Table 6. Properties of sepiolites.

Number	Condition	T/°C	AV/mPa·s	PV/mPa·s	YP/Pa	Gel/Pa/Pa	FL_{API}/mL
①	Before aging	65	3.00	2.00	1.00	0/0	—
		25	3.00	2.50	0.50	0/0	
		4	4.00	3.00	1.00	0/0	
	After aging	65	2.25	2.00	0.25	0/0	—
		25	2.75	2.00	0.75	0/0	
		4	4.00	3.50	0.50	0/0	
②	Before aging	65	3.50	3.00	0.50	0/0	—
		25	3.50	3.00	0.50	0/0	
		4	3.50	3.00	0.50	0/0	
	After aging	65	3.00	3.00	0	0/0	—
		25	3.25	3.25	0	0/0	
		4	3.75	3.50	0.25	0/0	

The data in Table 6 indicate that the sepiolite-based slurries exhibit low viscosity with minimal changes in properties before and after aging. The increase in apparent and plastic viscosity is not significant with decreasing temperature. The primary mineral composition of the sepiolite used in the experiment includes quartz, dolomite, and palygorskite, although the content of palygorskite is relatively low, suggesting a lower grade of the mineral source. Sepiolite is a type of hydrous chain silicate mineral containing aluminum and magnesium, typically exhibiting a fibrous crystal structure. When dispersed in a solution through stirring, the fibers interlock to form a chaotic network resembling a "tangled straw pile", which is crucial for the stable rheological properties of the sepiolite-based slurry. The mechanical parameters of the sepiolite fiber structure, rather than electrostatic interactions between particles, predominantly determine the rheological characteristics of the slurry. As the temperature decreases, the kinetic energy of the sepiolite fibers decreases, causing the originally dispersed fibers to aggregate into fiber bundles. This increases the internal friction between fibers, between fibers and fiber bundles, and between fiber bundles, leading to an increase in slurry viscosity. Additionally, the reduced kinetic energy enhances the rigidity of the sepiolite fibers, forming a resilient network structure within the dispersion system. This structure constrains free water, increasing the system's flow resistance and, consequently, the slurry viscosity. However, because the sepiolite fibers have a limited number of charged sites when dispersed in water, the electrostatic repulsion between the fibers is minimal. The resulting fibrous structure is stable, and the increase in slurry viscosity with decreasing temperature is therefore not pronounced.

(4) The performance of the two attapulgite-based slurries within the temperature range of 4°C to 65°C is presented in Table 7.

Table 7. Properties of palygorskites.

Number	Condition	T/°C	AV/mPa·s	PV/mPa·s	YP/Pa	Gel/Pa/Pa	FL_{API}/mL
①	Before aging	65	1.00	1.00	0	0/0	—
		25	1.50	1.00	0.50	0/0	
		4	2.00	1.50	0.50	0/0	
	After aging	65	1.00	1.00	0	0/0	—
		25	1.50	1.00	0.50	0/0	
		4	2.00	1.50	0.50	0/0	
②	Before aging	65	1.00	1.00	0	0/0	—
		25	1.50	1.50	0	0/0	
		4	2.25	2.00	0.25	0/0	
	After aging	65	1.00	1.00	0	0/0	—
		25	1.50	1.50	0	0/0	
		4	2.00	2.00	0	0/0	

The data in Table 7 reveal that the attapulgite-based slurries have a very low viscosity with insignificant changes in properties before and after aging. The increase in apparent and plastic viscosity with decreasing temperature is not pronounced. The primary mineral composition of the attapulgite used in the experiment consists of clay minerals, with the presence of quartz, potassium feldspar, and dolomite as impurities. Attapulgite shares a similar mineral morphology with sepiolite, characterized by a limited number of charged sites on the fiber surface due to hydrolysis of broken bonds. This results in a low electrostatic repulsion between attapulgite fibers, leading to a stable fibrous network structure and consistent rheological properties regardless of temperature changes. In addition to being temperature-insensitive, attapulgite and sepiolite-based slurries also exhibit excellent salt resistance. This is attributed to the fact that the surface charge of the crystals in solution primarily originates from the hydrolysis of broken bonds. However, both minerals have a low degree of isomorphism, resulting in a lack of structural charge and a low overall charge density. Consequently, the electrostatic repulsion between fiber bundles in water is significantly smaller than van der Waals forces, allowing attapulgite and sepiolite-based slurries to maintain their original rheological properties even at high salinities.

(5) The performance of the two kaolinite-based slurries within the temperature range of 4°C to 65°C is presented in Table 8.

The data in Table 8 indicate that the kaolinite-based slurries exhibit very low viscosity with minimal changes in properties before and after aging. The increase in apparent and plastic viscosity with decreasing temperature is not pronounced. The kaolinite samples primarily consist of clay minerals, with additional impurities such as quartz, plagioclase, and calcite. Kaolinite is classified as a 1:1 type clay mineral, with its crystal cell composed of one layer of oxygen-aluminum sheets and one layer of silicon-oxygen sheets. The OH layer and O layer are arranged opposite to each other, and the strong polarity of the OH bond readily forms hydrogen bonds between the crystal layers, causing these layers to bond tightly together. The interlayer spacing is small, resulting in a rigid crystal structure with almost no expandability, dispersibility, or swelling properties. There is also negligible lattice substitution, leading to a very low cation exchange capacity. Water molecules find it difficult to penetrate between the crystal layers, categorizing kaolinite as a non-swelling clay mineral. Consequently, kaolinite has poor hydration capacity and poor slurry-forming properties, with its rheological behavior being largely unaffected by external factors.

Table 8. Properties of kaolinites.

Number	Condition	T/°C	AV/mPa·s	PV/mPa·s	YP/Pa	Gel/Pa/Pa	FL_{API}/mL
①	Before aging	65	0.50	0.50	0	0/0	—
		25	1.50	1.50	0	0/0	
		4	2.00	2.00	0	0/0	
	After aging	65	0.50	0.50	0	0/0	—
		25	1.00	1.00	0	0/0	
		4	2.00	2.00	0	0/0	
②	Before aging	65	0.50	0.50	0	0/0	—
		25	1.50	1.50	0	0/0	
		4	2.00	2.00	0	0/0	
	After aging	65	0.50	0.50	0	0/0	—
		25	1.75	1.75	0	0/0	
		4	2.00	2.00	0	0/0	

4 CONCLUSION

Although the rheological properties of calcium bentonite, sepiolite, attapulgite, and kaolinite do not vary significantly with temperature, their poor slurry-forming performance and low base slurry viscosity (with apparent and plastic viscosities less than 4 mPa·s) render them unsuitable for providing the necessary functions such as viscosity enhancement, filtration reduction, and borehole and wall improvement during drilling operations. Consequently, they are not suitable as base slurry materials for deepwater marine water-based drilling fluids. Sodium bentonite-based slurries, on the other hand, exhibit moderate viscosity and their rheological properties change within a controllable range with temperature. Therefore, sodium bentonite (sample ③) is selected as the base slurry material for deepwater marine water-based drilling fluids. To ensure a constant rheological behavior within the temperature range of 4°C to 65°C subsequent experiments will systematically study the rheological properties of the sodium bentonite-based slurry.

REFERENCES

Arkadiusz Jo zefczak, Andrzej Skumiel. (2011). Ultrasonic investigation of magnetic nanoparticles suspension with PEG Biocompatible coating[J]. *Journal of Magnetism and Magnetic Materials*, (323): 1509–1516.

Basim Abu-Jdayil. (2011). Rheology of sodium and calcium bentonite–water dispersions: Effect of electrolytes and aging time[J]. *International Journal of Mineral Processing*, 98: 208–213.

Hanyi Zhong, Zhengsong Qiu, Weian Huang, *et al.* (2011). Shale inhibitive properties of polyether diamine in water-based drilling fluid[J]. *Journal of Petroleum Science and Engineering*, 78: 510–515.

Hege E, Majeed Y, Eirik S. (2001). Hydrate Control During Deepwater Drilling: Overview and New Drilling-Fluids Formulations[J]. *Spe Drilling & Completion*, 16(1): 19–26.

Rafael Roa, Felix Carrique, Emilio Ruiz-Reina. (2012). Ion size effects on the electrokinetics of salt-free concentrated suspensions in ac fields[J]. *Journal of Colloid and Interface Science*, (387): 153–161.

Robert Schlemmer, Jonathan Sheldon. (2010). Flat-rheology mud for deepwater wells[J]. *World Oil*, 231(1): 45–51.

Shuo Wu, Tian Xia, Naike Wang, *et al.* (2024). Phase diagram analysis of SBS modified bitumen using rheological and optical techniques[J]. *Journal of Polymer Research*, 32(1): 16–16.

Xin Zhang, Yu Zhao, Tianyi Zhang, *et al.* (2023). Potential of hydrolyzed wheat protein in soy-based meat analogues: Rheological, textural and functional properties[J]. *Food Chemistry: X*, 20100921–100921.

Young Gun Ko, Ung Su Choi, Yong Jin Chun. (2009). Influence of particle size on shear behavior of amine-group-immobilized polyacrylonitrile dispersed suspension under electric field[J]. *Journal of Colloid and Interface Science*, (335): 183–188.

Fluid Mechanics and Industrial Applications – Fang (Ed.)

Research on harmonic balance method for turbulent simulation of compressor flow

Ziwei Wang, Bin Li, Fan Liu, Weilong Zhang and Shuai Ma
China Aerodynamics Research and Development Center, Mianyang, Sichuan, China

ABSTRACT: The rotor/stator interference effect is one of the main sources of unsteady flow inside the compressor. The harmonic balance method can simulate the unsteady effects of rotor/stator interference based on the single passage grid, which is more efficient than the full annulus unsteady method. However, current research on harmonic balance technology focuses on the N-S equation, with very little involvement in harmonic balance calculations for turbulence models, especially transition models. This article implements harmonic balance calculations for the SA turbulence model, SST turbulence model, and $\gamma - \mathrm{Re}_\theta$ transition model, and conducts tests based on a single stage rotating cascade. The conclusion shows that for these three turbulence simulation methods, the harmonic balance calculation results are in good agreement with the results of the full annulus unsteady simulation.

1 INTRODUCTION

The compressor is the core component of an aircraft engine, and its aerodynamic performance directly affects the engine's thrust and efficiency. The compressor is composed of stator components and high-speed rotating rotor components, and the rotor/stator interference effect generated by the relative motion of the two is one of the main sources of the unsteady flow inside the compressor. The simulation of unsteady flow with rotor/stator interference generally requires unsteady calculations based on the full annulus grid. Based on the harmonic balance method, a single passage grid can be used to simulate the unsteady flow of rotor/stator interference, greatly reducing the computational complexity. Therefore, it is widely used in the flow simulation of compressors (Ekici, K. and Hall, K. C. 2007).

However, current research on harmonic balance technology focuses on N-S equations and rarely involves harmonic balance calculations for turbulence models, especially transition models.

This article will implement harmonic balance calculations for commonly used turbulence calculation models in compressor flow simulation, including SA (Spalart, P. and Allmaras, S.A. 1992) turbulence model, SST (Menter, F. 1993) turbulence models and $\gamma - \mathrm{Re}_\theta$ transition model (Menter, F. *et al.* 2006). Based on a compressor case for testing, the results of the full annulus unsteady simulation are used as a benchmark for verification.

2 NUMERICAL METHOD

2.1 *Harmonic balance governing equations for N-S equations*

For turbomachinery, the relative motion between the rotor and stator results in unsteady flow, which exhibits periodicity in time. In the absence of blade vibration and vortex shedding, the frequency sequence of a certain blade row flow is (Ekici, K. and Hall, K.C. 2010.)

DOI: 10.1201/9781003679561-10

$$\omega_k^{rowi} = \sum_{j=1, j \neq i}^{nRow} n_{k,j} \cdot \left| B_j(\Omega_j - \Omega_i) \right| \tag{1}$$

Among them, the subscript *nRow* is the total number of blade rows in the turbomachinery, B_j is the number of blades in the j-th row, the rotational speed of the i-th row is Ω_i, and $n_{k,j}$ is any non-negative integer. When $n_{k,j}$ is taken as 1, ω_k^{rowi} is the fundamental frequency of the disturbance caused by blade row j in the blade row *rowi*.

The governing equation of the harmonic balance method has been modified based on the N-S equation, and the derivation process is as follows. The finite volume form of the N-S equation is

$$\int_{\Omega} \frac{\partial U}{\partial t} dV + \oint_{\partial\Omega} \vec{F} \cdot \vec{n} ds = 0 \tag{2}$$

When not considering changes in grid volume, its semi discrete form is as follows

$$V \frac{\partial U}{\partial t} + R(U) = 0 \tag{3}$$

The governing equation of the harmonic balance method is derived based on the N-S equation, and the specific process is as follows.

Firstly, the periodic spatial flow field variables are expanded into Fourier series in the time direction, and the first N orders of the series are truncated as follows

$$\begin{aligned} U &= \overline{U} + \sum_{n=1}^{N} U_{cn} \cos(n\omega t) + U_{sn} \sin(n\omega t) \\ R &= \overline{R} + \sum_{n=1}^{N} R_{cn} \cos(n\omega t) + R_{sn} \sin(n\omega t) \end{aligned} \tag{4}$$

In the formula, N is the harmonic order, which can be set to any non-negative integer, and ω is the fundamental frequency of periodic flow. $\bar{U}$ is the mean value, U_{cn} and U_{sn} are the coefficients of each order of the series, independent of t, and can be expressed as

$$\begin{aligned} \widetilde{U} &= \left[\overline{U}, U_{c1}, \cdots, U_{s1}, \cdots\right]^T \\ \widetilde{R} &= \left[\overline{R}, R_{c1}, \cdots, R_{s1}, \cdots\right]^T \end{aligned} \tag{5}$$

Substituting Equation (5) into Equation (4), it can be expressed as

$$\begin{aligned} U &= [1, \cos(\omega t), \cdots, \sin(\omega t), \cdots] \cdot \widetilde{U} \\ R &= [1, \cos(\omega t), \cdots, \sin(\omega t), \cdots] \cdot \widetilde{R} \end{aligned} \tag{6}$$

Substituting Equation (4) into Equation (3) and merging similar terms, there are

$$(0 + \bar{R}) + (\omega V U_{s1} + R_{c1})\cos(\omega t) + (-\omega V U_{c1} + R_{s1})\sin(\omega t) + \cdots = 0 \tag{7}$$

The above equation holds for any t, then the coefficients of all trigonometric terms are 0

$$\begin{aligned} &\left(0 + \overline{R}\right) = 0 \\ &(\omega V U_{s1} + R_{c1}) = 0 \\ &(-\omega V U_{c1} + R_{s1}) = 0 \\ &\vdots \\ &(N\omega V U_{sN} + R_{cN}) = 0 \\ &(-N\omega V U_{cN} + R_{sN}) = 0 \end{aligned} \tag{8}$$

Substituting Equation (5) into Equation (8), there is

$$VM\widetilde{U} + \widetilde{R} = 0 \tag{9}$$

In the formula, M is a $2N$+1-order square matrix, and its expression is as follows

$$M = diag(0, -\omega, -2\omega, \cdots, -N\omega, \omega, 2\omega, \cdots, N\omega) \tag{10}$$

Divide a flow period into 2N+1 equal parts, take the discrete quantity in time, that is, the physical quantity of each time layer, and express it as follows

$$\begin{aligned} U^* &= [U(t_0 + \Delta t), \cdots, U(t_0 + (2N+1)\Delta t)]^T \\ R^* &= [R(t_0 + \Delta t), \cdots, R(t_0 + (2N+1)\Delta t)]^T \end{aligned} \tag{11}$$

The relationship between the original variable and the Fourier coefficients can be obtained from Equation (6), that is

$$\begin{aligned} U^* &= E^{-1}\widetilde{U} \\ R^* &= E^{-1}\widetilde{R} \end{aligned} \tag{12}$$

The expression for E^{-1} is as follows

$$E^{-1} = \begin{bmatrix} 1 & \cos(\omega t_1) & \cdots & \cos(N\omega t_1) & \sin(\omega t_1) & \cdots & \sin(N\omega t_1) \\ 1 & \cos(\omega t_2) & \cdots & \cos(N\omega t_2) & \sin(\omega t_2) & \cdots & \sin(N\omega t_2) \\ \vdots & \vdots & \vdots & \vdots & \vdots & \vdots & \vdots \\ 1 & \cos(\omega t_{2N+1}) & \cdots & \cos(N\omega t_{2N+1}) & \sin(\omega t_{2N+1}) & \cdots & \sin(N\omega t_{2N+1}) \end{bmatrix} \tag{13}$$

In Equation (13), there are

$$t_i = t_0 + i \cdot \Delta t \tag{14}$$

Substituting Equation (12) into Equation (9), there is

$$VMEU^* + ER^* = 0 \tag{15}$$

Multiplying the equation by E^{-1} on the left yields the governing equation for harmonic balance method

$$V(E^{-1}ME)U^* + R(U^*) = 0 \tag{16}$$

Among them, $E = (E^{-1})^{-1}$ is the discrete Fourier transform matrix. At this point, the unsteady equation system has been transformed into a steady equation system at 2N+1 time steps.

Using the pseudo time advancement method, solve this steady-state equation system in a synchronous time advancement manner

$$V\frac{\partial U^*}{\partial \tau} + R(U^*) + V(E^{-1}ME)U^* = 0 \tag{17}$$

2.2 *Harmonic balance governing equations for turbulence models*

2.2.1 *SA model*

The governing equation of the SA model is as follows

$$\begin{aligned} \frac{\partial \tilde{v}}{\partial t} + u_j\frac{\partial \tilde{v}}{\partial x_j} = C_{b1}[1 - f_{t2}]\tilde{S}\tilde{v} + \frac{M_\infty}{\mathrm{Re}}\left\{C_{b1}[(1 - f_{t2})f_{v2} + f_{t2}]\frac{1}{\kappa^2} - C_{w1}f_w\right\}\left(\frac{\tilde{v}}{d}\right)^2 \\ - \frac{M_\infty}{\mathrm{Re}}\frac{C_{b2}}{\sigma}\tilde{v}\frac{\partial^2 \tilde{v}}{\partial x_j^2} + \frac{M_\infty}{\mathrm{Re}}\frac{1}{\sigma}\frac{\partial}{\partial x_j}\left[(v + (1 + C_{b2})\tilde{v})\frac{\partial \tilde{v}}{\partial x_j}\right] \end{aligned} \tag{18}$$

Similar to the N-S governing equation of harmonic balance transformation, establishing a governing equation system of 2N+1 SA models. The working variables of the SA model on 2N+1 time layers are expressed as follows

$$\widetilde{v}^* = [\widetilde{v}(t_0 + \Delta t), \cdots, \widetilde{v}(t_0 + (2N+1)\Delta t)]^T \tag{19}$$

Convert the time partial derivative of the working variables in the SA model into

$$\frac{\partial \widetilde{v}}{\partial t} \rightarrow (E^{-1}ME)\widetilde{v}^* \tag{20}$$

In this way, the harmonic balance calculation of the SA model can be achieved.

2.2.2 *SST model*

The SST model was proposed by Menter and consists of transport equations for turbulent kinetic energy k and turbulent dissipation rate ω. The time partial derivative terms in the two transport equations are $\frac{\partial(\rho k)}{\partial t}$ and $\frac{\partial(\rho\omega)}{\partial t}$, respectively, where ρ is the density.

Similarly, establish a governing equation system for 2N+1 transition models. The transition variables on 2N+1 time layers are expressed as follows

$$\begin{aligned}(\rho k)^* &= [(\rho k)(t_0 + \Delta t), \cdots, (\rho k)(t_0 + (2N+1)\Delta t)]^T \\ (\rho\omega)^* &= [(\rho\omega)(t_0 + \Delta t), \cdots, (\rho\omega)(t_0 + (2N+1)\Delta t)]^T\end{aligned} \tag{21}$$

The time partial derivatives of the transition variables are transformed into

$$\begin{aligned}\frac{\partial(\rho k)}{\partial t} &\rightarrow (E^{-1}ME)(\rho k)^* \\ \frac{\partial(\rho\omega)}{\partial t} &\rightarrow (E^{-1}ME)(\rho\omega)^*\end{aligned} \tag{22}$$

In this way, harmonic balance calculation of SST turbulence model can be achieved.

2.2.3 $\gamma - \mathrm{Re}_\theta$ *Transition model*

The $\gamma - \mathrm{Re}_\theta$ transition model was proposed by Menter and Langtry in 2004. This model needs to be combined with the SST turbulence model, and the role of the transition model is ultimately achieved through the eddy viscosity coefficient.

The $\gamma - \mathrm{Re}_\theta$ transition model is based on the SST turbulence model and supplemented with two transport equations for transition prediction (Du, Y.F. *et al.* 2023): one is the intermittent factor γ, and the other is the local transition onset momentum thickness $\widetilde{Re}_{\theta t}$. The time partial derivative terms in the two transport equations are $\frac{\partial(\rho\gamma)}{\partial t}$ and $\frac{\partial(\rho\widetilde{Re}_{\theta t})}{\partial t}$.

Similarly, establish a governing equation system for 2N+1 transition models. The transition variables on 2N+1 time layers are expressed as follows

$$\begin{aligned}(\rho\gamma)^* &= [(\rho\gamma)(t_0 + \Delta t), \cdots, (\rho\gamma)(t_0 + (2N+1)\Delta t)]^T \\ \left(\rho\widetilde{\mathrm{Re}}_{\theta t}\right)^* &= \left[\left(\rho\widetilde{\mathrm{Re}}_{\theta t}\right)(t_0 + \Delta t), \cdots, \left(\rho\widetilde{\mathrm{Re}}_{\theta t}\right)(t_0 + (2N+1)\Delta t)\right]^T\end{aligned} \tag{23}$$

The time partial derivatives of the transition variables are transformed into

$$\begin{aligned}\frac{\partial(\rho\gamma)}{\partial t} &\rightarrow (E^{-1}ME)(\rho\gamma)^* \\ \frac{\partial\left(\rho\widetilde{\mathrm{Re}}_{\theta t}\right)}{\partial t} &\rightarrow (E^{-1}ME)\left(\rho\widetilde{\mathrm{Re}}_{\theta t}\right)^*\end{aligned} \tag{24}$$

In this way, the harmonic balance calculation of the $\gamma - \mathrm{Re}_\theta$ transition model can be achieved.

3 TEST CASES, RESULTS AND DISCUSSIONS

3.1 *Single passage case*

The single passage rotating cascade grid is shown in Figure 1. The first row is the rotor, with 36 blades. The second row is stator, with 46 blades. The number of grid cells is 25.3 thousand, the number of grid points is 35.3 thousand, and 5 grid points are arranged in the spanwise direction.

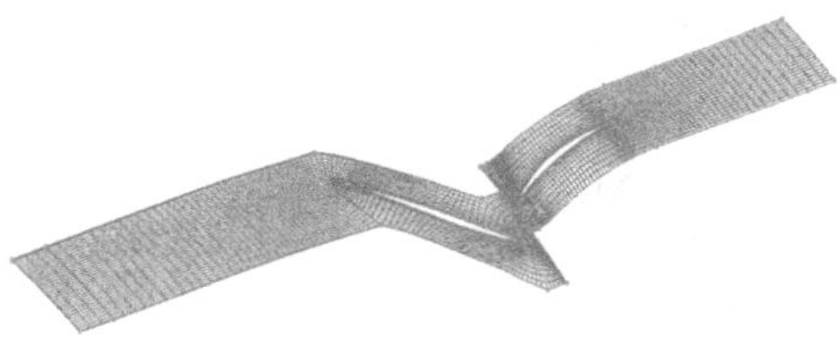

Figure 1. Single passage grid.

Mixing plane calculation and harmonic balance calculation based on single passage grid.

3.3 *Full annulus case*

The full annulus rotating cascade grid is obtained by rotating and replicating the single passage grid, as shown in Figure 2. The number of grid cells is 1.0278 million, and the number of grid points is 1.3980 million.

Figure 2. Full annulus grid.

Perform full annulus unsteady calculations based on the full annulus grid.

3.3 *Simulation setting*

For all cases, the rotor speed is -17188.7 rpm, the inlet total temperature is 101325 Pa, the inlet total pressure is 288.15 K, the inlet direction of velocity is axial, and the outlet is given a static pressure of 101325 Pa.

For mixing plane calculations, the single passage grid is used, spatial periodicity is applied at periodic boundaries, and mixing average is used at blade row interfaces for steady-state calculations.

For harmonic balance calculation, the single passage grid is used, and phase lag periodic processing is applied at the periodic boundary. Harmonic reconstruction is used for interpolation at the blade row interface to perform 5th order harmonic balance calculation.

For the full annulus unsteady calculation, the full annulus grid is used without periodic boundaries. Dynamic interpolation is performed at the blade row interface for unsteady calculation, and 2000 physical time steps are set within one rotation of the rotor.

3.4 *Results and discussions*

The variation in outlet mass flow is shown in Figure 3. The T in the horizontal axis represents the period of unsteady flow in the stator computational domain, which is equal to the time it takes for the rotor to rotate one rotor blade passage. The mixing plane method is a steady calculation method, so the results do not change over time.

In the following figures, ‘SA’ represents the SA turbulence model, ‘SST’ represents the SST turbulence model, and ‘Transition’ represents the $\gamma - Re_\theta$ transition model.

It can be seen that the mixing plane method cannot capture the unsteady effects of rotor/stator interference. Under the three turbulence models, the outlet mass flow variation calculated by harmonic balance is in good agreement with the full annulus unsteady simulation.

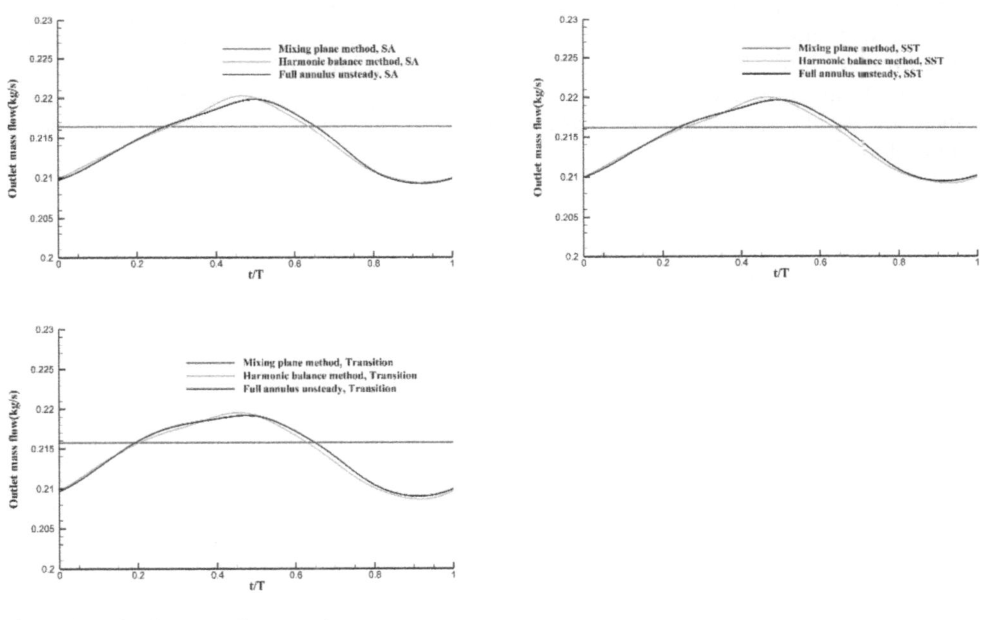

Figure 3. Outlet mass flow results.

The total pressure ratio is an important indicator of the aerodynamic performance of a compressor, and its variation is shown in Figure 4. It can be seen that the mixing plane method cannot capture the unsteady effects of rotor/stator interference. Under the three turbulence models, the total pressure ratio variation calculated by harmonic balance is in good agreement with the full annulus unsteady simulation.

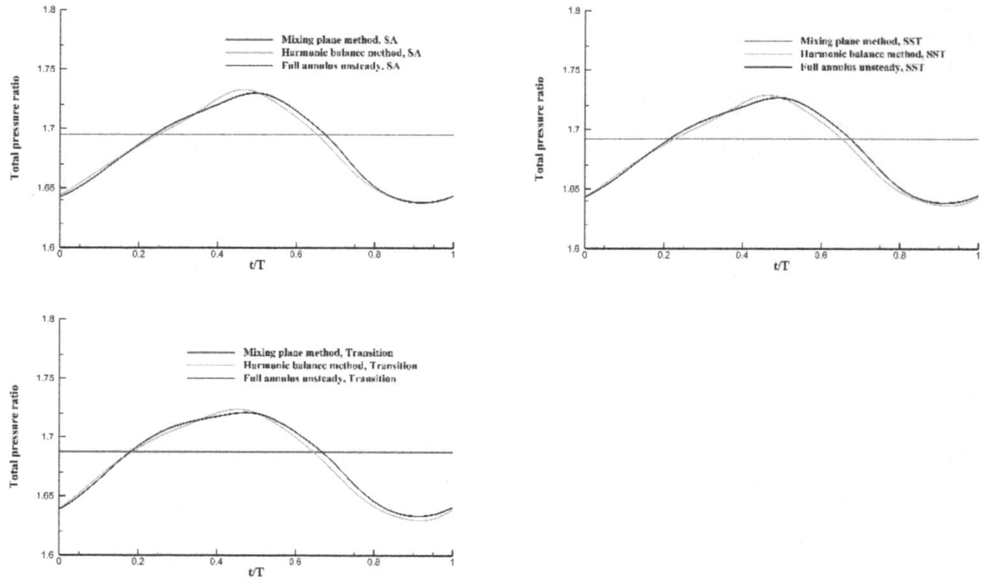

Figure 4. Total pressure ratio results.

Adiabatic efficiency is an important indicator of the aerodynamic performance of a compressor, and its variation is shown in Figure 5. It can be seen that the degree of agreement between adiabatic efficiency is slightly worse than that between outlet mass flow and total pressure ratio. Because adiabatic efficiency is calculated by combining total pressure ratio and total temperature ratio, resulting in error superposition. At this point, the adiabatic efficiency calculated by the mixing plane method has deviated from the full annulus unsteady simulation.

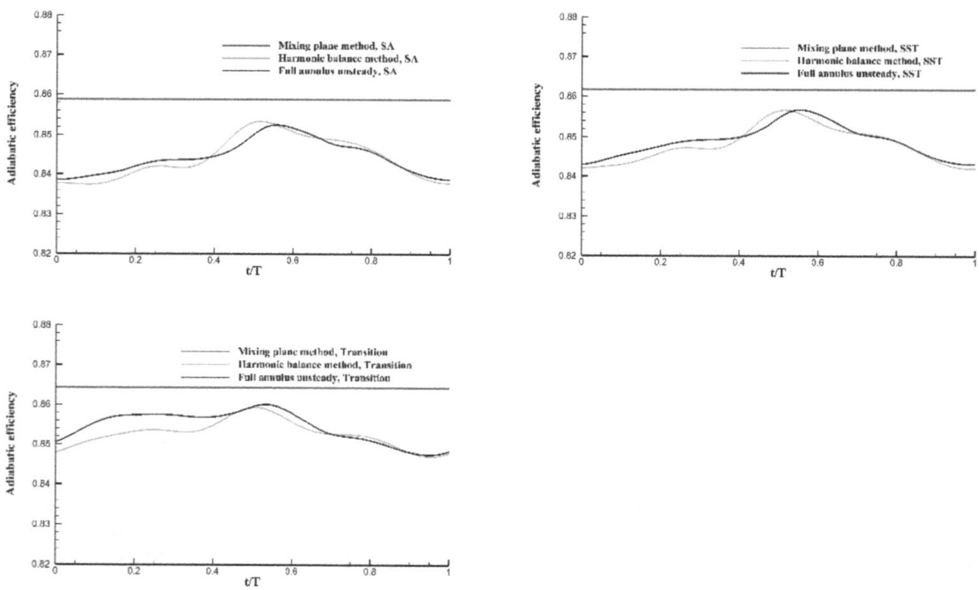

Figure 5. Adiabatic efficiency results.

4 CONCLUSIONS

Derived harmonic balance calculation formulas for SA turbulence model, SST turbulence model, and $\gamma - \mathrm{Re}_\theta$ transition model. Tests were conducted based on a single stage rotating cascade, and the results were compared with those of the mixing plane method and the full annulus unsteady method. The following conclusions were drawn:

(a) Under the three turbulence simulation methods, the harmonic balance calculation results are in good agreement with the full annulus unsteady simulation, indicating that the harmonic balance method based on a single passage grid effectively captures the unsteady effects of rotor/stator interference.
(b) Compared to the single passage steady state method (mixing plane method), the harmonic balance method is also based on a single passage grid but can obtain results that are much closer to the full annulus unsteady simulation.

ACKNOWLEDGMENT

This work was supported by the Natural Science Foundation of China under grant no. 12402275.

REFERENCES

Du, Y.F. Cao, Y.H. Zhou, L. *et al.* (2023). Preliminary study on hypersonic correction of r-re transition model based on fluent UDF. *Journal of North University of China (Natural Science Edition)*. 44(3).

Ekici, K. Hall, K.C. (2010). Harmonic balance analysis of blade row interactions in a transonic compressor. *Journal of Propulsion and Power*. 26(2): 335–343.

Ekici, K. and Hall, K.C. (2007). Nonlinear analysis of unsteady flows in multistage turbomachines using harmonic. *AIAA Journal.* 45(5):1047–1057.

Menter, F. (1993). Zonal two equation turbulence models for aerodynamic flows. *AIAA Paper*. 1993–2906.

Menter, F. Langtry, R.B. Likki, S.R. (2006). A correlation-based transition model using local variables — part I: model formulation. *Journal of Turbomachinery*. 413–422.

Spalart, P. and Allmaras, S.A. (1992). A one-equation turbulence model for aerodynamic flows. *AIAA Paper*. 1992–0439.

Fluid Mechanics and Industrial Applications – Fang (Ed.)
© 2026 The Author, ISBN: 978-1-041-15371-9

Energy efficiency analysis of heated tobacco products using coupled numerical models of heating device and cigarette

Zhe Xiong
China Tobacco Hubei Industrial Co., Ltd., Wuhan, China

Shuo Wang
School of Power and Mechanical Engineering, Wuhan University, Wuhan, China

Lei Liu, Junjie Hong, Qiuxiang Shi, Ruoxi Li, Yiming Zhang, Guchun Li, Xin Hao, Ting Huang and Xi Xiao
China Tobacco Hubei Industrial Co., Ltd., Wuhan, China

ABSTRACT: The widespread application of heated tobacco products (HTPs) faces numerous challenges: low thermal energy utilization, limited device runtime, and localized surface overheating. Addressing these issues requires a systematic understanding of energy dynamics across the entire heating tobacco system. To address this gap, this study introduces a comprehensive modeling framework that integrates heat and mass transfer processes across both the heating device and the cigarette. The model is validated through complementary experimental techniques: thermo-gravimetric analysis to characterize tobacco pyrolysis kinetics, thermocouple-based temperature profiling and infrared thermography to map surface heat distribution. Based on this model, we propose a novel method for evaluating system energy efficiency. Numerical simulations reveal energy distribution dynamics during operation, elucidating the correlation between energy efficiency and puffing frequency, as well as quantifying the temporal evolution of heat losses across key components. The findings reveal the internal thermal characteristics of the coupled system and provide practical guidance for the design and optimization of next-generation products.

1 INTRODUCTION

New tobacco products demonstrates vast market potential and promising development prospects, numerous organizations and companies have intensified research and development efforts toward novel tobacco products, such as e-cigarette and heated tobacco products (HTPs) (Erin, 2013; Ratajczak A, 2020). HTPs use external heat sources to warm the tobacco rather than igniting (Daher N, 2010; Titz B, 2020), thereby reducing the generation of harmful compounds associated with pyrolysis and combustion (Eaton D, 2018; Hashizume T, 2023; Jiang Z, 2018; Smith M R, 2016). It has become one of the mainstream alternatives to traditional cigarettes. However, due to structural limitations or suboptimal material choices, the low thermal efficiency of current heating products leads to increasingly prominent issues with insufficient operational duration. As the global market for HTPs expands rapidly, addressing key challenges such as low system efficiency and uneven temperature distribution has become critical to advancing technological innovation within the industry.

DOI: 10.1201/9781003679561-11

In recent years, many research institutions worldwide have investigated and optimized the heat transfer processes within heating products and tobacco. Sha *et al.* proposed a novel perforated sidewall structure for cigarettes that lowers aerosol temperature by enhancing the cooling mechanism. Their simulations demonstrated that this design reduces the formation of harmful substances in the smoke (Ding S, 2024). Luo *et al.* further analyzed the independent effects of the sidewall perforations on airflow and heat transfer, providing a theoretical basis for optimizing the size and placement of openings (Luo J, 2025). In heating tobacco products, researchers also focus on improving the release efficiency of key components like nicotine while minimizing the emission of other toxic substances. Zhang *et al.* developed a numerical model to study heat transfer and component release in tobacco. Their simulations revealed that factors such as component ratios, tobacco rod diameter, heating rate, and preheating time influence aerosol composition, offering guidance for tobacco product design (Zhang B, 2024). Kim *et al.* designed a dual-source heating system incorporating both needle and wall heaters. By clarifying the interplay between internal heat transfer and pyrolysis reactions, they achieved a 4.6-fold increase in nicotine release efficiency (Kim S H, 2022). Due to the complex nature of substance release in tobacco, Nordlund *et al.* developed a coupled numerical model of the tobacco and the heating device to simulate puff-by-puff release of glycerol and nicotine, and validated the results using gas chromatography–mass spectrometry (GC–MS) measurements (Nordlund M, 2016). Current thermal analyses of heating tobacco system are largely focused on localized regions, lacking a comprehensive modeling and optimization approach for the entire heat transfer system. Traditional studies tend to address thermal issues by focusing on temperature distributions themselves, rather than analyzing the heating system from the fundamental perspective of energy allocation. Moreover, evaluations of thermal efficiency from an energy performance standpoint remain insufficient. To address these limitations, this study proposes a coupled model for heating tobacco system, emphasizing energy efficiency analysis to inform the structural design of heating products.

This study develops a coupled heat transfer model for the pin-style heating tobacco products (most popular), including both the heating device and cigarette, and performs numerical simulations of the overall heat and mass transfer processes. We propose a comprehensive energy efficiency evaluation framework, the total input energy in the system is defined as the heat generated by the heating needle. In the flow and heat transfer model of the cigarette, we model tobacco and filter sections as porous media, with internal heat transfer assumed to be in thermal equilibrium (Alomar O R, 2017). The release of chemical components is quantitatively described by the Arrhenius-based reaction mode (Zhang B, 2024). We simulated the global heat and mass transfer behavior using ANSYS Fluent and validated the model accuracy through TG analysis and temperature measurements. This work provides a robust foundation for designing advanced heating products with strong market competitiveness.

2 NUMERICAL MODELLING METHODS

2.1 *Physical model*

This study uses the MOK FWRD center-pin-style heated device and specially designed "COO" cigarette purchased from the heated tobacco market as shown in Figure 1. Structural dimensions were measured physically with an accuracy of 0.1 mm. The cigarette comprises four main parts: tobacco section, hollow section, filter section, and wrapping paper. The cigarette has a diameter of 7.16 mm and consists of a 12 mm tobacco section, a 26 mm hollow section, and a 7 mm filter section.

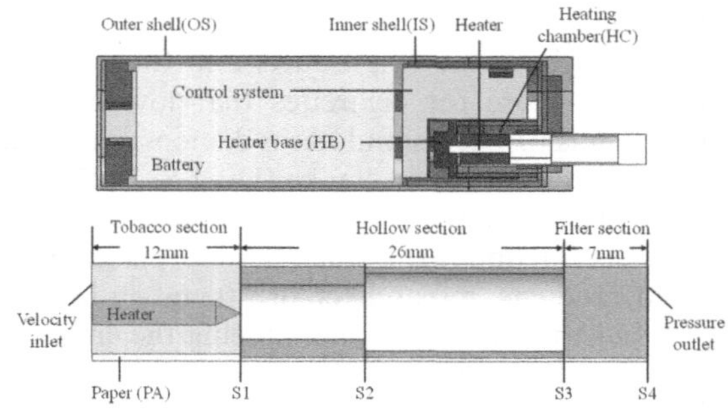

Figure 1. Structural diagram of "MOK FWRD" heating product and "COO" cigarette.

The heater consists of internal and outer shell, battery, heating chamber, control system, heating needle, and needle base. The cylindrical heating needle has a diameter of 2 mm and inserts 12 mm into the tobacco section. Due to its non-axisymmetric geometry and internal electronic complexity, detailed modeling of the device is impractical for heat and mass transfer simulations. Therefore, the model is simplified to retain only the main structure centered on the heating needle. During operation, the ceramic needle reaches approximately 300 °C. Air enters from the bottom of the cigarette and exits through the filter, which traps most aerosol particles.

2.2 *Governing equation*

The physical processes during the operation of the system are as follows: conduction from the heating needle to the tobacco section, porous medium heat transfer within the reconstituted tobacco sheets, convective heat exchange between the airflow and tobacco sheets, and conduction through device components. Air enters the cigarette from the bottom and exits through the filter end. The governing equations for mass, momentum, and energy conservation are as follows:

$$\frac{\partial \rho_g}{\partial t} + \nabla \cdot \left(\rho_g \vec{v}\right) = 0 \tag{1}$$

$$\frac{\partial}{\partial t}\left(\rho_g \vec{v}\right) + \nabla \cdot \left(\rho_g \vec{v}\vec{v}\right) = -\nabla P + \nabla \cdot \left[\mu\left(\nabla \vec{v} + \nabla \vec{v}^{T}\right)\right] + \rho_g \vec{g} + S_i \tag{2}$$

$$\frac{\partial}{\partial t}(\rho E) + \nabla\left(\vec{v}(\rho E + P)\right) = S_h + \nabla \cdot \left[k\nabla T + \left(\vec{\tau} \cdot \vec{v}\right)\right] \tag{3}$$

Where ρ_g is gas density, t represents time, $\vec{v}$ is gas flow rate, P represents pressure, μ is gas dynamic viscosity of, $\vec{g}$ represents gravitational acceleration, S_i is momentum source term, E represents the total energy generated during the working cycle, T is temperature, k represents the thermal conductivity, $\vec{\tau}$ is viscous stress tensor, and S_h represents the energy term. According to the theory proposed by J. Pascal and H. Pascal (Pascal, J P, 1995, 1997), the low Reynolds number airflow in the tobacco justifies the use of Darcy's law to calculate flow resistance in the porous media. During puffing, the pressure drop is primarily attributed to viscous resistance; however, as airflow velocity increases, inertial effects become significant, further reducing pressure. Assuming porous media materials in both tobacco and filter sections are isotropic, the momentum source term is simplified as (Nield D, 2020):

$$S_i = -\left(\frac{\mu}{a}\vec{v} + C_2\frac{1}{2}\rho\left|\vec{v}\right|\vec{v}\right) \tag{4}$$

Where μ is air dynamic viscosity, α represents permeability. C_2 is inertia coefficient, which represents the pressure drop per unit length in the air flow direction.

2.3 *Kinetics model for thermal decomposition of tobacco*

The tobacco combustion model used in this study is proposed by Ali Rostami (Rostami A, 2003), with kinetic parameters sourced from earlier literature (Chen S, 2020; Kim S H, 2022; Sun Z, 2017). (Zhu H, 2024) Tobacco undergoes three distinct stages during heating: moisture evaporation, pyrolysis, and oxidative combustion. This section focuses on the release of nicotine and glycerol during evaporation and pyrolysis, modeled using first-order Arrhenius kinetics:

$$\frac{dY_i}{dt} = -A_i e^{-\frac{E_{a,i}}{RT}} Y_i \tag{5}$$

Where Y_i is the mass fraction of component I, A_i represents the pre-exponential factor, $E_{a,i}$ represents the activation energy of component I, R is the universal gas constant, T is the tobacco temperature. The reaction rate of evaporation of liquid water into water vapor is described by Equation (6) (Chen S, 2020):

$$\frac{d\rho_w}{dt} = -A_w e^{-\frac{E_w}{RT}} \cdot \left(-\frac{\rho_w}{\rho_{w0}}\right)^{N_{ri}} \cdot \rho_{w0} \tag{6}$$

Where ρ_w is mass concentration of water in tobacco, ρ_{ro} is the initial mass concentration of water, which is 74kg/m^3, T_s is the solid surface temperature, and N_w is the reaction order. The release of nicotine and glycerol is controlled by two independent pyrolysis precursors, and the pyrolysis reaction is expressed as (Chen S, 2020)

$$\frac{\partial \rho_{ri}}{\partial t} = -A_{ri} e^{-\frac{E_{ri}}{RT_s}} \cdot \left(-\frac{\rho_{ri}}{\rho_{r0}}\right)^{N_{ri}} \cdot \rho_{r0} \tag{7}$$

$$\frac{\partial \rho_r}{\partial t} = \sum_{i=1}^{4} \frac{\partial \rho_{ri}}{\partial t} \tag{8}$$

Where ρ_{ri} is the mass concentration of the ith pyrolysis precursor, ρ_r is the total mass concentration of the pyrolysis precursor, ρ_{ro} is the initial mass concentration of the pyrolysis precursor, which is 596.8kg/m^3.

2.4 *Material and boundary conditions*

Parameters such as tobacco density and initial porosity were experimentally measured, while other values were referenced from previous studies (Zhang B, 2024). We simulate heat and mass transfer of the coupled system by ANSYS Fluent, with the following assumptions made to reduce computational complexity (Baker R R, 2004; Wójtowicz M A, 2023):(1) The gas flow within the porous medium is assumed to be steady and laminar throughout the simulation;(2) A local thermal equilibrium is assumed between the gas phase and the solid matrix in the porous medium;(3) The gas is considered incompressible within the porous medium, and the solid skeleton of the porous structure is assumed to be isotropic;(4) The gas flow within the porous medium obeys the ideal gas law;(5) The thermal properties of the porous solid matrix are assumed to be independent of temperature;(6) The gas flow is assumed to occur solely in the z-direction, corresponding to the puffing direction. Puffing conditions followed the Health Canada Intense (HCI) regimen: the heating needle preheats for 30 s, followed by a 2 s puff of 55 mL every 30 s. Both velocity inlet and pressure outlet during puffing were modeled with periodic boundary conditions controlled via a user-defined function (UDF). The heating power was measured experimentally and set to 1 W in simulations. Second-order upwind schemes were used for pressure, momentum, and energy discretization, with first-order implicit time stepping. Pressure–velocity coupling was handled via the coupled scheme. Natural convection boundary conditions were applied at air interfaces, with heat transfer coefficients determined from material properties and temperature gradients. Grid independence was verified based on surface temperature distribution, and a grid with 7.8 million cells was ultimately used for simulation.

3 RESULT AND DISCUSSION

3.1 *Thermal decomposition and component release in tobacco*

We simulated the tobacco heating process to evaluate the release of three key components in each puff, then verified the model using thermos-gravimetric analysis (TGA) on ground tobacco sheets powder under a nitrogen atmosphere at a heating rate of 10 K/min. By comparing the simulated total mass loss curve with TG curve, we confirmed the model's accuracy.

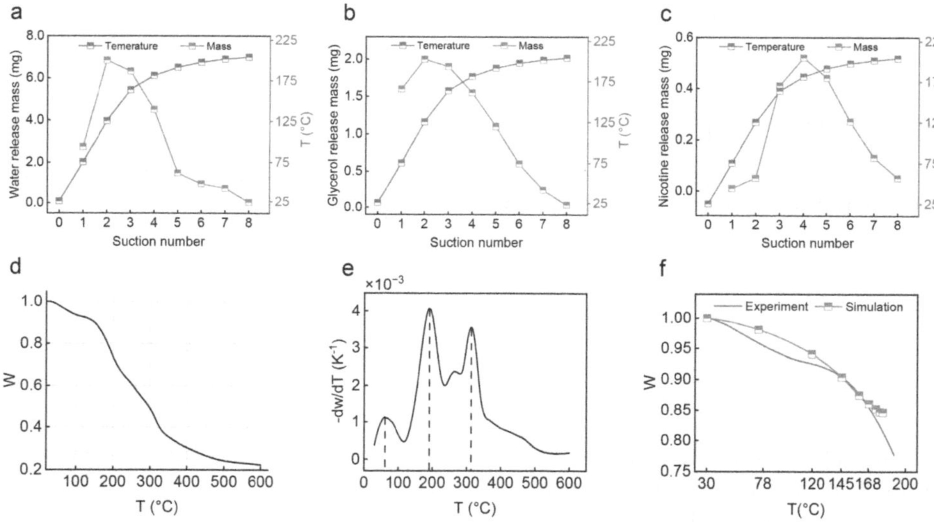

Figure 2. (a), (b), (c) Release amounts of water vapor, glycerin, and nicotine per puff and the corresponding temperatures, respectively. (d) TG curve (w represents mass fraction). (e) DTG curve. (f) comparison of simulated weight loss curve and TGA curve.

Figure 2a, b, c shows the simulated puff-by-puff release of water vapor, glycerol, and nicotine, along with the average tobacco temperature during each puff. During the first puff, the average tobacco temperature reaches around 75 °C. However, the area near the center-positioned heating needle heats more rapidly, which triggers water vapor release even in the initial puff. Figure 2d shows the TGA-derived tobacco mass loss curve. We calculated its derivative to obtain the mass loss rate curve shown in Figure 2e. Our simulations indicate that the heating system raises the average tobacco temperature to no more than 200 °C. Therefore, we focus on the TGA results before the second mass loss peak. Figure 2f illustrates that the simulated and experimental mass loss curves show overall agreement below 200 °C, supporting the accuracy of the release model. Most aerosol components released during heating are absorbed in the hollow and filter sections, which affects internal heat transfer.

3.2 *Heat transfer of the cigarette during puffing*

To understand the internal thermal behavior of the cigarette under realistic operating conditions, we simulate the transient temperature field during puffing and compare the results with experimental measurements using thermocouples. Following the HCI puffing regime, after preheating the tobacco for 30 seconds, we drew a steady 55 mL puff over 2 seconds. A puff was applied every 30 seconds, and the process was considered complete after the eighth puff at 151 seconds. Figure 3a shows the cross-sectional temperature field of the cigarette. During preheating, the temperature exhibits a uniform gradient along the puffing direction. Once puffing begins, the center region heats up more rapidly than the sidewalls. After puffing ends, the central temperature drops sharply. Figure 3b displays the temperature profiles at cross-sections S1–S4 during puffing. All four planes show the highest temperature at the center and the lowest near the edges, consistent with the characteristics of center-needle-type heating. Figure 3c illustrates the peak temperatures at these sections. Under preheating, S1 reaches about 270 °C, S2 about 165 °C, and S3 about 60 °C, while S4, the surface exposed to ambient air, shows almost no increase. The outlet surface temperature of the cigarette stick was measured using a needle-type thermocouple and compared to the simulation results. Figure 3d shows strong agreement between the two, with a maximum deviation of 6 °C, confirming the accuracy of the heat transfer model. Together with the previous section, this study verifies the internal heat and mass transfer model of the cigarette. Since the system efficiency analysis requires modeling of the coupled system of cigarette and heating device, the next section focuses on simulation and analysis of coupled system.

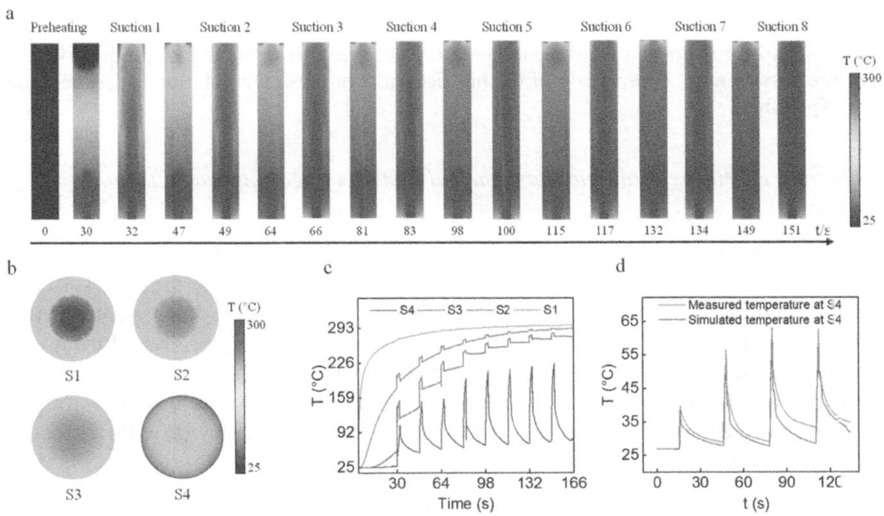

Figure 3. (a) Temperature distribution and changes inside the cigarette. (b) temperature distribution of S1-S4 section (The cross section corresponding to S1-S4 is shown in Figure 1). (c)diagram of temperature change for S1-S4. (d)simulated and experimental values of outlet temperature.

3.3 *Heat transfer analysis for coupled system of heating products and tobacco*

Each puffing cycle lasts 300 seconds, and a new cigarette is inserted at the end of the cycle. Natural convection occurs between the device surface and the surrounding air, and the heat transfer coefficient varies with temperature as given in Equation (9):

$$q = \frac{C \cdot k}{L} \cdot \left(\frac{g\beta L^3}{\nu\alpha}\right)^{0.25} \cdot (T - T_{air})^{1.25} \tag{9}$$

where C is the dimensionless empirical coefficient of heat flux density, k is the thermal conductivity of air, L is the characteristic length, β is the thermal expansion coefficient, v is the fluid dynamic viscosity, α is the air thermal diffusivity, T is the heat exchange surface temperature. The surface temperature variation of the device obtained from simulation is shown in Figure 4b. The corresponding experimental results captured using an infrared thermal camera are presented in Figure 4a. The spatial distribution of hot spots confirms the accuracy of the heat transfer model. Figure 4c compares the maximum surface temperatures of the device during the puffing cycle. The maximum deviation between simulation and experiment is less than 3 °C, confirming that the model is suitable for energy efficiency analysis and calculations.

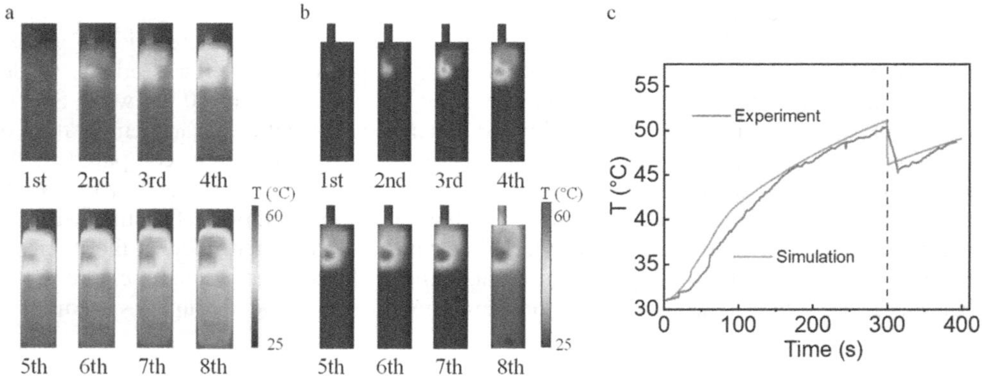

Figure 4. (a) Infrared image of surface temperature changes. (b) simulated graph of surface temperature variation. (c) comparison of the highest temperatures of simulation experiment and infrared experiment.

3.4 *Thermal efficiency evaluation for coupled system of heating products and tobacco*

The portion of total heat generated by the heating needle that is used to heat the tobacco is defined as effective heat. The ratio of effective heat to total heat defines the system energy efficiency. The heat carried by the high-temperature aerosol during puffing contributes to the effective heat, but this component is difficult to quantify. Energy losses during operation include the heat absorbed by device components (Q_a) and the convective loss to ambient air from the device surface (Q_c), both of which can be readily estimated. Q_c is obtained by integrating the total heat flux over the convective surface of the device with respect to time, Q_a is determined from the temperature rise of each component:

$$Q_a = \sum_{i=1} c_i \rho_i V_i \cdot (T_i - T_0) \tag{10}$$

This section evaluates how system energy efficiency evolves with puff number by calculating the total and lost heat during each heating cycle. Each puffing cycle lasts 300 seconds, with the heating needle operating at a constant power of 1 W. We performed a numerical simulation for six consecutive cigarettes to reflect different puffing frequency. The simulation yielded cumulative heat loss, cumulative energy efficiency, and per-cycle values over six puffing cycles. As shown in Figure 5a and 5b, during the first cycle, most heat loss results from internal component heating, while the low device surface temperature leads to convective losses below 10%. As puff count increases, the heating rate of components decreases, leading to a downward trend in component heat absorption per cycle. Meanwhile, the rising surface temperature causes an increasing trend in convective heat loss. When puffing three or more cigarettes continuously, convective loss dominates the heat loss in each cycle. Figure 5c illustrates the cumulative energy efficiency, which starts at 21% in the first cycle and increases with puff number, stabilizing at 48% after six cycles. Figure 5d presents the energy efficiency for each cycle: 21% in the first cycle, a significant rise to 44.9% in the second, 57.7% in the fifth, and 58.6% in the sixth. The 0.9% difference between the last two indicates that system efficiency stabilizes after the fifth puffing cycle.

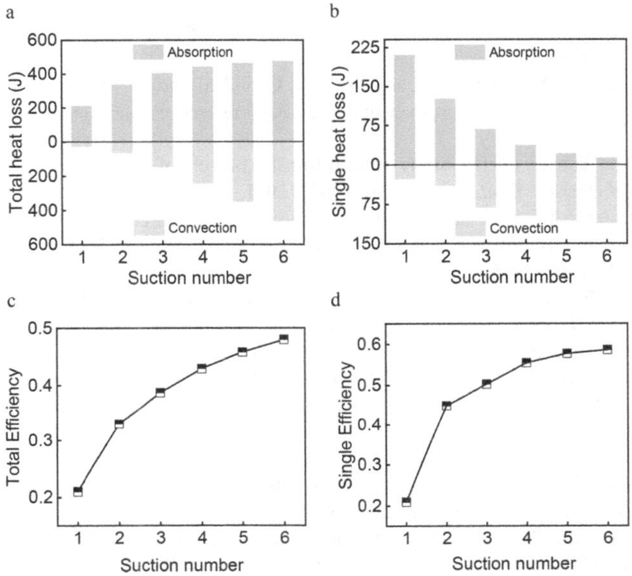

Figure 5. (a) Diagram of total heat absorption and dissipation. (b)diagram of heat absorption and dissipation in each cycle. (c) diagram of cumulative efficiency for the coupled system; (d) diagram of efficiency in each cycle.

4 CONCLUSION

To address the problems of low energy efficiency, insufficient device endurance, and localized overheating in heating tobacco products, this study proposes to analyze the heat and mass transfer behavior from the perspective of energy efficiency, based on coupled modeling of heating device and cigarette. The conclusions are as follows: (1) A coupled heat transfer model of coupled system was established, and the overall heat transport process was analyzed through numerical simulation. (2) A definition of system energy efficiency was proposed as the proportion of heating needle output that is effectively utilized; the relationship between heat loss and puffing frequency was obtained through numerical simulation.

REFERENCES

Alomar O R, Mendes M A A, Trimis D, *et al.* (2017). Numerical simulation of complete liquid–vapour phase change process inside porous media: a comparison between local thermal equilibrium and non-equilibrium models[J]. *International Journal of Thermal Sciences*, 112: 222–241.

Baker R R, Bishop L J. (2004). The pyrolysis of tobacco ingredients[J]. *Journal of Analytical and Applied Pyrolysis*, 71(1): 223–311.

Chen S, Liu H, Sun Z, *et al.* (2020). Numerical simulation of the distribution of in situ cigarette combustion-generated particulate matter[J]. *Colloids and Interfaces*, 4(4): 59.

Daher N, Saleh R, Jaroudi E, *et al.* (2010). Comparison of carcinogen, carbon monoxide, and ultrafine particle emissions from *narghile* waterpipe and cigarette smoking: sidestream smoke measurements and assessment of second-hand smoke emission factors[J]. *Atmospheric Environment*, 2010, 44(1): 8–14.
Ding S, Zhang J, Tuo S, *et al.* (2024). Filling structure design and cooling mechanism study of the heat-not-burn cigarettes with sidewall openings[J]. *International Communications in Heat and Mass Transfer*, 153: 107405.
Eaton D, Jakaj B, Forster M, *et al.* (2018). Assessment of tobacco heating product THP1.0. Part 2: product design, operation and thermophysical characterisation[J]. *Regulatory Toxicology and Pharmacology*, 2018, 93: 4–13.
Erin L. Sutfin, Thomas P. McCoy, Holly E.R. Morrell, *et al.* (2013). Electronic cigarette use by college students[J]. *Drug and Alcohol Dependence*, 131(3): 214–221.
Hashizume T, Ishikawa S, Matsumura K, *et al.* (2023). Chemical and in vitro toxicological comparison of emissions from a heated tobacco product and the 1R6F reference cigarette[J]. *Toxicology Reports*, 10: 281–292.
Jiang Z, Ding X, Fang T, *et al.* (2018). Study on heat transfer process of a heat not burn tobacco product flow field[J]. *Journal of Physics: Conference Series*, 1064: 12011.
Kim S H, Choi H, Jung Y, *et al.* (2022). Numerical analysis of the coupling between heat transfer and pyrolysis in heat-not-burn tobacco using computational fluid dynamics[J]. *Korean Journal of Chemical Engineering*, 39(11): 2907–2915.
Luo J, Zhang Y, Qi Y P. (2025). The influence of sidewall holes on gas flow and heat transfer inside the heat-not-burn cigarette[J]. *International Communications in Heat and Mass Transfer*, 164: 108949.
Nield D A. (2000). Resolution of a paradox involving viscous dissipation and nonlinear drag in a porous medium[J]. *Transport in Porous Media*, 2000, 41(3): 349–357.
Nordlund M, Kuczaj A. (2016). Modeling flow, heat and mass transfer in a porous biomass plug - when used in an electrically heated tobacco system[C]. *Sixth International Conference on Porous Media and Its Applications in Science, Engineering and Industry.*
Pascal J P, Pascal H. (1997). Non-linear effects on some unsteady non-darcian flows through porous media[J]. *International Journal of Non-Linear Mechanics*, 32(2): 361–376.
Pascal, J P, Pascal, *et al.* (1995). Pressure diffusion in unsteady non-darcian flows through porous media (journal article) | ETDEWEB[J]. *European Journal of Mechanics. B, Fluids*, 14(1): 75–90.
Ratajczak A, Jankowski P, Strus P, *et al.* (2020). Heat not burn tobacco product—a new global trend: impact of heat-not-burn tobacco products on public health, a systematic review[J]. *International Journal of Environmental Research and Public Health*, 17(2): 409.
Rostami A, Murthy J, Hajaligol M. (2003). Modeling of a smoldering cigarette[J]. *Journal of Analytical and Applied Pyrolysis*, 66(1): 281–301.
Smith M R, Clark B, Lüdicke F, *et al.* (2016). Evaluation of the tobacco heating system 2.2. Part 1: description of the system and the scientific assessment program[J]. *Regulatory Toxicology and Pharmacology*, 81: S17–S26.
Sun Z, Wen J, Luo X, *et al.* (2017). An improved CFD model of gas flow and particle interception in a fiber material[J]. *Chinese Journal of Chemical Engineering*, 25(3): 264–273.
Titz B, Szostak J, Sewer A, *et al.* (2020). Multi-omics systems toxicology study of mouse lung assessing the effects of aerosols from two heat-not-burn tobacco products and cigarette smoke[J]. *Computational and Structural Biotechnology Journal*, 18: 1056–1073.
Wójtowicz M A, Bassilakis R, Smith W W, *et al.* (2003). Modeling the evolution of volatile species during tobacco pyrolysis[J]. *Journal of Analytical and Applied Pyrolysis*, 66(1): 235–261.
Zhang B, Xiao L, Huang J, *et al.* (2024). Numerical study on heat transfer and release characteristics of key components in electrically heated tobacco products[J]. *International Journal of Chemical Engineering*, 2024 (1): 6690487.
Zhu H, Wu C, Qin Y, *et al.* (2024). Construction of a numerical model for cigarette smoking and combustion and simulation of combustion cone shape[J]. *Fluid Dynamics Research*, 2024, 56(6): 65501.

Author index

For Product Safety Concerns and Information please contact our EU representative GPSR@taylorandfrancis.com Taylor & Francis Verlag GmbH, Kaufingerstraße 24, 80331 München, Germany